TO
VICKI, SONDRA, AND JON

Introduction

to

Physical Electronics

enc

ES W

Introduction

to

Physical Electronics

———

KARL E. LONNGREN

Department of Electrical and Computer Engineering
The University of Iowa
Iowa City

Allyn and Bacon, Inc.

Boston   London   Sydney   Toronto

Library of Congress Cataloging-in-Publication Data

Lonngren, Karl E. (Karl Erik), 1938–
     Introduction to physical electronics/Karl E. Lonngren.
          p.   cm.
     Bibliography: p.
     Includes index.
     ISBN 0-205-11141-6
 ·   ISBN 0-205-11409-1 (*International*)
1. Quantum electronics.   2. Solid state electronics.
3. Electrodynamics.   4. Plasma (Ionized gases)   I. Title.
QC688.L66 1988
537.5––dc19   87-21491
                                                  CIP

Printed in the United States of America

10 9 8 7 6 5 4 3 2 1   92 91 90 89 88 87

Contents

Chapter Four

Energy Levels   55

Chapter Five

Semiconductors   71

Chapter Six

pn Junctions   95

Preface

The investigation of the electrical properties of materials that can be used as elements in modern circuits is an important subject for study in the final decades of the twentieth century and beyond. In addition to having a certain degree of intrinsic interest to a small parochial community of scholars, these materials also form the basis for the worldwide growth industry of modern society. The words "high tech" appear in print in several contexts but usually within the spirit of describing the industry of the future. As one might expect, the path that leads from the scientific laboratory to the commercial marketplace can have many detours that may be imposed by certain laws of nature. These cannot be violated, even with the best of intentions of a well-meaning entrepreneur. It is to this end that we have formulated a textbook that summarizes these basic laws, which were developed prior to and during the depths of the worldwide economic depression, and then elucidates their application to the modern-day development of the myriad electronic elements that have been sent into the heavens aboard rockets, have been weaseled under the earth in mechanical moles, and have recently been inserted into the human body to prolong life.

The first third of this book is dedicated to the exposition of the need for the development of quantum mechanics in the first third of this century and some of the ramifications of its predictions. This will include a discussion of the required statistical tools needed to interpret these predictions. It will then be possible to understand some of the important electrical properties of materials from what are considered to be "first principles." This will lead into the topic of semiconductors, both those found in nature and those developed in the laboratory.

The second third of the book describes how one combines these materials in various configurations to form the electronic circuit elements that have revolutionized the last half of the twentieth century. These elements include diodes and the several forms of transistors, the acronyms of which appear to have been taken from a can of alphabet soup.

The final third of this book examines the fundamentals of devices that are either new or are based on a nonsolid-state foundation. The devices may owe their origin to the motion of single particles or to their collective interaction. This latter topic will lead into a description of high-frequency and high-power tubes, where solid-state principles have not yet been found to be applicable, and into some fundamentals of plasma physics.

The philosophy of this text is to develop all the necessary physical concepts from the basic laws of nature so that the reader can obtain physical insight into the operation of a particular device. The development requires an understanding of the elements of mechanics, thermodynamics, and electrostatics, such as one would obtain in the first year of an undergraduate physics course, and a mathematical maturity that would normally be found at that level.

This text is written for juniors or seniors in Electrical and Computer Engineering who wish to gain a knowledge of how the devices that have revolutionized modern electronics actually work. This course would typically follow an introductory electronic circuits course, where the student is just exposed to "black boxes." This text would also be useful for the same level student in physics who desires to gain an understanding of where the physical principles that are being acquired can be applied. Rather than present a collection of numbers that are either out of date or are quickly forgotten, the text tries to develop just the principles and leaves the numbers for the homework problems, and more appropriately, for the reader's later professional life.

The author wishes to acknowledge Professor David Andersen, Dr. Edward Gabl, Dr. Hulbert Hsuan, and Professor James Nordman for their suggestions concerning the presentation of the material in this text and Professor Sudhakar Reddy and Dean Robert Hering for their support of this work. He also thanks his family for their encouragement and understanding during this endeavor and the book is dedicated to them. This work was supported in part by the National Science Foundation.

The author extends special thanks to the following reviewers whose contributions have enriched the text:

Professor Stanley Burns
The Iowa State University

Professor Daniel Leenov
Lehigh University

Professor Alwyn C. Scott
The University of Arizona

K.E.L.

Introduction

to

Physical Electronics

Chapter One

Elements of Quantum Mechanics

The knowledge of the physical behavior of nature that was possessed by mankind at the end of the nineteenth century seemed adequate to explain the many observed phenomena that confronted the scientists and engineers of the day. The equations developed by Isaac Newton and James Clerk Maxwell were rich and described the many features of nature that were the everyday experiences of these people as they took their walks in the woods and contemplated such esoteric subjects as the heavenly motion of planets, leaves falling to the ground, waves crashing on beaches, and lightning emanating from clouds. These equations were even nonlinear, so large-amplitude phenomena could be described. Hence the motion of a swing at large angles of deviation or a transformer near saturation when inserted in an electrical circuit seemed treatable by known equations and techniques. It was a quiet time among these learned people since it appeared that all that was to be discovered that could be considered new and fundamental had already been discovered and that nature's breadbasket of new basic truths was empty. The idea that there might be some relation between waves and particles seemed not to be a topic worthy of discussion as scholars rustled amid the fall colors in the woods. Even in science, the tranquility of autumn had entered. Little did they know of the revolution that was forthcoming in the basic laws of physics, which would eventually lead to the development of the alphabet soup of solid-state devices that have had such an impact in the last half of the twentieth century.

During the early part of the twentieth century, certain experiments were performed whose results could not be interpreted in terms of the classical physics that was known at the time. Some of these experiments will be reviewed here so that the reader can sense some of the turmoil that scientists faced after the turn of the century. These experiments led to the development of **quantum mechanics**, which we will then introduce. Many of the giants who pioneered this work and have led us through the dark forest have important effects, equations, and constants named after them. These will be noted as the reader progresses through this development.

I. Waves

The reader is probably well acquainted with particle motion, so only a brief review of the properties of waves and the wave equation will be presented. This will allow the reader to understand the Schrödinger wave equation, which is fundamental for solid-state materials. The reader who already understands waves can easily move on to the next section, which is concerned with the experiments that led to the development of quantum mechanics, with no loss of continuity.

The propagation of a wave ψ can be described by a set of first-order partial differential equations:

$$\frac{\partial \psi}{\partial t} + c \frac{\partial \psi}{\partial x} = 0 \qquad \textbf{(a)}$$

$$\frac{\partial \psi}{\partial t} - c \frac{\partial \psi}{\partial x} = 0 \qquad \textbf{(b)}$$

$$\textbf{(1)}$$

These first-order equations can also be combined and written as one second-order partial differential equation:

$$\frac{\partial^2 \psi}{\partial t^2} - c^2 \frac{\partial^2 \psi}{\partial x^2} = 0 \qquad \textbf{(2)}$$

In both equations (1) and (2), c corresponds to the velocity of propagation, x is the spatial coordinate, and t is time. The most general solution of (2) is

$$\psi(x, t) = \psi_R(x - ct) + \psi_L(x + ct) \qquad \textbf{(3)}$$

where ψ_R and ψ_L correspond to the waves that propagate to the right and left, respectively. The functions ψ_R and ψ_L are solutions of (1a) and (1b), respectively, as can also be verified by direct substitution. For example, let $\theta = x - ct$ and substitute ψ_R in (1a). We find

$$\frac{\partial \psi_R}{\partial t} + c \frac{\partial \psi_R}{\partial x} = \frac{d\psi_R}{d\theta}\left(\frac{\partial \theta}{\partial t}\right) + c\frac{d\psi_R}{d\theta}\left(\frac{\partial \theta}{\partial x}\right) = 0$$

where the chain rule has been applied.

Example 1.1   Given that an initial excitation at $t = 0$ of a wave pulse that will propagate to the right is given by

$$\psi_R(x, 0) = 1 - |x|, \qquad |x| \le 1$$
$$= 0 \qquad\qquad |x| \ge 1$$

sketch its behavior as time increases. Let the velocity $c = 1$.

Answer: The solution is $\psi_R(x, t) = 1 - |x - ct|$, where $c = 1$. This is valid for $|x - ct| \le 1$; otherwise $\psi_R = 0$. Its behavior is shown in the following sketch for three times.

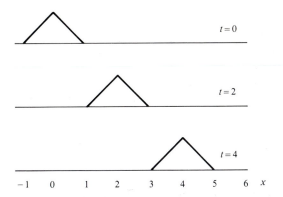

The physical interpretation for the functions ψ_R and ψ_L are that they are quantities that move to values of increasing and decreasing values of position x as the time t increases (no one is getting younger in age as they read this). Remember that to evaluate a function of a dependent variable a number must be specified! Try and find $\sin x$ on your calculator without specifying a number for the variable x. The function will be the same if x and t adjust themselves such that the number stays the same. Therefore if $\theta = x - ct$ has a value, say 3, then x must increase as t increases in order to obtain the same function. The wave has moved to the right. A similar argument holds for the wave moving to the left. Equation (3) includes both components of the waves.

Let us now examine the behavior of the wave response that propagates away from a sinusoidal excitation at $x = 0$; that is,

$$\psi(x = 0, t) = \psi_0 \sin(-\omega t)$$

The frequency of oscillation is f hertz (Hz), and the radian frequency $\omega = 2\pi f$. It is common to also define the frequency with the Greek letter v, and this will also be used here. Let us assume that the excited wave propagates only to the right. Therefore, one can write

$$\psi(x, t > 0) = \psi_0 \sin(kx - \omega t)$$

where k is called the wave number and $k = 2\pi/\lambda$, where λ is the wavelength of the wave. This can be rewritten as

$$\psi(x, t > 0) = \psi_0 \sin\left[k\left(x - \frac{\omega}{k}t \right) \right]$$

If we compare this with equation (3), we recognize that the ratio ω/k has the same role as the velocity c. This velocity corresponds to the velocity of a point of constant phase of the wave; that is, $\Delta\theta = 0$, where

$$\psi = \psi_0 \sin \theta$$

It is therefore called the *phase velocity* of the wave and it is given by

$$c = \frac{\omega}{k} \tag{4}$$

Another velocity for waves will be encountered in the work that leads to the development of quantum mechanics and the Schrödinger equation; this velocity is called the *group velocity*, which can be derived as follows. Let there exist two propagating waves in the same region of space at the same time given by

$$\psi_a = \psi_{a0} \sin(k_a x - \omega_a t)$$

and

$$\psi_b = \psi_{b0} \sin(k_b x - \omega_b t)$$

Let the amplitudes of the two waves be the same, $\psi_{a0} = \psi_{b0} = \psi_0$. Also let the frequencies and wave numbers be slightly different than a mean value ω_0 and k_0 such that

$$\omega_a = \omega_0 + \Delta\omega \qquad k_a = k_0 + \Delta k$$

$$\omega_b = \omega_0 - \Delta\omega \qquad k_b = k_0 - \Delta k$$

Since the media in which these waves are propagating is assumed to be linear, we need only add these two signals together to obtain the total response.

$$\psi = \psi_a + \psi_b$$

$$\psi = \psi_{a0} \sin(k_a x - \omega_a t) + \psi_{b0} \sin(k_b x - \omega_b t)$$

$$= \psi_0 \{\sin[(k_0 + \Delta k)x - (\omega_0 + \Delta\omega)t]$$

$$+ \sin[(k_0 - \Delta k)x - (\omega_0 - \Delta\omega)t]\} \tag{5}$$

$$\psi = 2\psi_0 \cos[\Delta k x - \Delta\omega t] \cdot \sin[k_0 x - \omega_0 t]$$

where a trig identity has been employed. Equation (5) states that the two signals will "phase mix" together and act as an amplitude-modulated wave. This phase mixing is shown in Figure 1.1. The individual sine wave will propagate at the phase velocity given by (4). However, a value of constant amplitude (i.e., the envelope) will propagate with the group velocity, which is defined as

$$v_g = \lim_{\Delta k \to 0} \left(\frac{\Delta\omega}{\Delta k}\right) = \frac{\partial\omega}{\partial k} \tag{6}$$

A medium that has a nonlinear relation between the frequency and the wave number is called a *dispersive medium*. Examples include electromagnetic waveguides, electromagnetic waves propagating in an ionized gas called a plasma, or optical fibers.

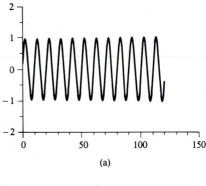

(a)

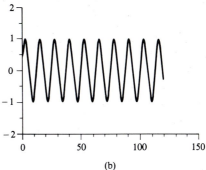

(b)

Figure 1.1   Phase mixing of two sine waves that differ in frequency by 25%. (a) $f = f_0$, (b) $f = 0.8f_0$, (c) the sum of the two sine waves.

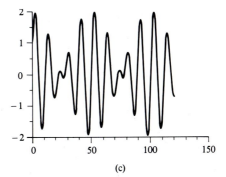

(c)

Example 1.2   It is possible to approximate the dispersion relation for certain types of media as $\omega = \omega_0 \sin k/k_0$. Calculate the phase and group velocities at $k/k_0 = 0$ and $k/k_0 = \pi/2$.

Answer: First derive the two velocities.

$$v_{\text{phase}} = \frac{\omega}{k} = \frac{\omega_0 \sin(k/k_0)}{k}$$

$$v_{\text{group}} = \frac{\partial \omega}{\partial k} = \frac{\omega_0 \cos(k/k_0)}{k_0}$$

Therefore

k/k_0	v_{phase}	v_{group}
0	ω_0/k_0	ω_0/k_0
$\pi/2$	ω_0/k	0

Instead of the wave propagating in an infinite media, let us insert large plane surfaces at $x = 0$ and at $x = L$, where $L > 0$, as shown in Figure 1.2. Several different boundary conditions may apply for different planes. For example, $\partial\psi/\partial x = 0$ or $\psi = 0$ at these surfaces may be valid boundary conditions for different planes. We will assume that $\psi = 0$ at the planes. In addition, we will specify the functional form of the wave to be of the form

$$\psi = Ae^{i(kx-\omega t)} + Be^{-i(kx+\omega t)}$$
$$= \{Ae^{ikx} + Be^{-ikx}\}e^{-i\omega t} \tag{7}$$

where A and B are arbitrary constants. Applying the boundary condition that $\psi = 0$ at $x = 0$, one finds that $B = -A$ and (7) can be written as

$$\psi = \{(-2i)A \sin kx\}e^{-i\omega t}$$

The constant A is specified by the amplitude of the initial excitation. Since $\psi = 0$ at $x = L$, k must satisfy

$$k = \frac{n\pi}{L} \tag{8}$$

One notes that there are **discrete** or **quantized** values of the integer n where this will occur. These discrete integers n are sometimes called eigenvalues or characteristic values. Recall that $k = \omega/c$ and hence $n\pi/L = \omega/c$. Therefore, only certain discrete frequencies will be allowed. This is similar to the discrete resonance frequency that one encounters in a circuit containing an inductor and a capacitor. The superposition of these two oppositely directed propagating waves creates a wave that stays in one location and is called a **standing wave**. In Figure 1.3, standing waves for three values of n are illustrated ($n = 1, 2,$ and 3). The concept of waves can be extended to three dimensions in a straightfor-

0   L

Figure 1.2 Two parallel plane surfaces at $x = 0$ and at $x = L$.

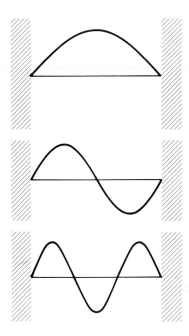

Figure 1.3   Three lowest modes of the standing wave.

ward manner. The more general wave equation corresponding to (2) is

$$\frac{\partial^2 \psi}{\partial t^2} - c^2 \nabla^2 \psi = 0 \tag{9}$$

where the Laplacian operator ∇^2 is

$$\nabla^2 = \frac{\partial^2}{\partial x^2} + \frac{\partial^2}{\partial y^2} + \frac{\partial^2}{\partial z^2}$$

in rectangular coordinates. A solution for (9) can be obtained using the technique of separation of variables, and this technique will be presented in Section III, where solutions of the Schrödinger equation are obtained. A general solution consists of waves that are propagating to increasing values and to decreasing values of the coordinates. Also, a more general condition for the allowed frequencies that involves the three coordinates will be found. This involves three numbers for the eigenvalues that generalize the value n given in (8), since now three dimensions are involved.

II. Early Experiments

A description of two experiments will be given to illustrate some of the troubles that were faced by the scientific community as they approached the time of the great worldwide economic depression. The first experiment was due to Franck and Hertz in 1914 and the second was due to Davisson and Germer in 1927. They clearly illustrate the **quantized** features of atoms and the duality of waves and particles.

To understand the Franck–Hertz experiment, consider the circuit in Figure 1.4. A small vacuum tube filled with low-pressure mercury vapor (a mercury vapor thyraton) is inserted in the circuit as shown. The cathode–grid structure acts as an electron gun, which sprays energetic electrons that are "boiled off" the cathode and accelerated by the voltage ϕ_0, which is monitored with the voltmeter into the region between the grid and the anode. The anode has a small negative bias applied to it, so only energetic electrons will be collected. The current I passing through the anode is monitored as a function of the voltage ϕ_0 that is applied between the cathode and the grid [i.e., the energy of the accelerated electrons that have a velocity $v = (2e\phi_0/m_e)^{1/2}$ after leaving the grid]. Instead of observing a straight line, a perturbed line with perturbations occurring at the same discrete value of $\Delta\phi = 4.86$ volts is detected, as indicated in Figure 1.5. In addition to measuring the voltage and the current, the emitted light is also monitored. It was found that the light has a wavelength of $\lambda = 2536$ Å (1 angstrom = 10^{-10} meter).

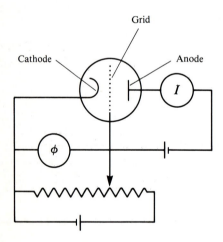

Figure 1.4 Schematic of the Franck–Hertz experiment.

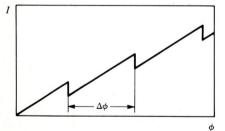

Figure 1.5 Measured current–voltage characteristics obtained in the Franck–Hertz experiment. The deviations $\Delta\phi$ were all equal to 4.86 volts.

The understanding of the atom at the time of the experiment could not totally explain the observed phenomena, although as we see in hindsight the understanding should have been obvious. The first serious model for an atom was due to Bohr, and it consisted of a model of a central core consisting of a positively charged nucleus around which circulated an electron. This describes accurately the hydrogen atom. See Figure 1.6.

An equilibrium is set up between the Coulomb force,

$$F_{\text{Coulomb}} = \frac{q^2}{4\pi\varepsilon_0 r^2}$$

which causes the particles to collide, and the centrifugal force,

$$F_{\text{centrifugal}} = \frac{mv^2}{r}$$

which causes the electron to quickly pass off to a great distance. Equating these two forces, we write

$$\frac{q^2}{4\pi\varepsilon_0 r^2} = \frac{mv^2}{r} \tag{10}$$

The energy of the atom consists of potential energy due to the Coulomb force and a kinetic energy due to the electron's being in motion. This is written as

$$\text{energy} = -\frac{q^2}{4\pi\varepsilon_0 r^2} + \frac{mv^2}{2} = -\frac{q^2}{8\pi\varepsilon_0 r} \tag{11}$$

where the potential energy at $r = \infty$ is taken to be zero. An electron would neither gain nor lose energy if it stayed in the same orbit. However, it would gain or lose energy if it changed from orbit 1, where its energy is E_1, to orbit 2, where its energy is E_2. This energy would either be absorbed or radiated. According to

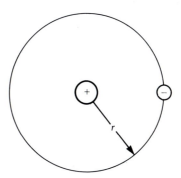

Figure 1.6   Bohr model of an atom.

Einstein, this energy is given by

$$E_2 - E_1 = hv = h\frac{c}{\lambda} \tag{12}$$

where h is Planck's constant ($h = 6.6 \times 10^{-34}$ joule-second and $\hbar = h/2\pi$). The frequency of the light is v and the wavelength λ of the light is given by $\lambda = c/v$, where c is the velocity of light ($c \approx 3 \times 10^8$ meters/second).

Bohr further quantized these orbits by asserting that the angular momentum of an electron in an orbit would be quantized and set

$$mvr = n\frac{h}{2\pi} = n\hbar \tag{13}$$

where n is an integer. Eliminate v and r in equations (10), (11), and (13) to write

$$E = -\frac{q^2}{8\pi\varepsilon_0 r} = -\frac{q^2}{8\pi\varepsilon_0\{[4\pi\varepsilon_0 m(rv)^2/q^2]\}} = -\frac{q^2}{8\pi\varepsilon_0[(4\pi\varepsilon_0 m)/q^2][n\hbar/2m]^2}$$

or

$$E_n = -\frac{q^4 m}{8n^2 h^2 \varepsilon_0^2} \tag{14}$$

These energy levels are **discrete**.

Let us compare these ideas with the Franck–Hertz experiment, where a discrete structure was noted in the current–voltage curve (Figure 1.4). To do this, we assume that the measured discrete voltage intervals (4.86 V) correspond to discrete energy jumps and that a photon of light is emitted. Hence from equation (12) we compute

$$4.86 \text{ eV} = h\frac{c}{\lambda_0}$$

$$\lambda_0 = \frac{(6.6 \times 10^{-34} \text{ joule-second})(3 \times 10^8 \text{ meters/second})}{(4.86 \text{ electron volts})(1.6 \times 10^{-19} \text{ coulomb})}$$

and find $\lambda_0 = 2551$ Å, which is in good agreement with the measured value of 2536 Å.

The second experiment that led to an understanding of the duality between waves and particles was the experiment of Davisson and Germer, which involved the scattering of electrons from a piece of nickel and explaining the measurements with a wave theory. To set the framework of the experiment, consider first Bragg scattering of a wave from uniformly distributed, perfectly reflecting solid spheres, as shown in Figure 1.7. If the distance $abc = a'b'c' + n\lambda$, where λ is the

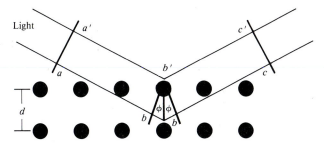

Figure 1.7   Bragg scattering of light.

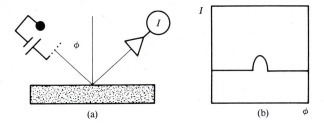

(a)   (b)

Figure 1.8   Measured detected current versus angle in the Davisson–Germer experiment.

wavelength of the incident wave and n is an integer, then a strong signal will be detected at the plane cc'. If d is the vertical distance between the spheres, then

$$a'b'c' = abc - n\lambda = abc - 2d \sin \phi$$

Therefore,

$$2d \sin \phi = n\lambda \tag{15}$$

which is known as Bragg's law.

The Davisson–Germer experiment scattered electrons from a nickel crystal that possesses a uniformly distributed crystal structure of scattering objects with a known separation d'. In performing their experiment as shown in Figure 1.8(a), they noted a peak in the detected signal at a particular angle ϕ_{DG} that seemed anomalous. This peak is shown in Figure 1.8(b). Certain features of the crystal are known or can be calculated, such as the distance between the atoms. This can be illustrated by an example.

Example 1.3   Calculate the distance d' between the scattering objects in a potassium chloride (KCl) crystal.

Answer: Assume that the crystal could be broken up into cubes as shown in the following figure. The density is ρ, the molecular weight of KCl is W, and each atom

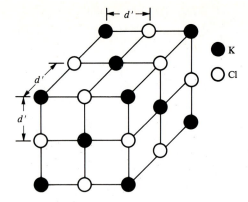

fills up a cube with a volume d'^3. The KCl molecule fills up a volume $2d'^3$. There are then $\frac{1}{2}d'^3$ molecules/meter^3 or $\frac{1}{2}\rho d'^3$ molecules/kilogram. From Avogadro's number $N_0(N_0 = 6.025 \times 10^{26})$, which tells the number of molecules per kilogram molecular weight, we write the number of molecules per kilogram as N_0/W;

$$\frac{N_0}{W} = \frac{1}{2\rho d'^3} \quad \text{or} \quad d' = \left[\frac{W}{2\rho N_0}\right]^{1/3}$$

For KCl, we compute

$$d' = \left[\frac{74.56}{2 \times 1990 \times 6.025 \times 10^{26}}\right]^{1/3} = 3.14 \text{ Å}$$

The molecular weight W is determined from chemical experiments.

From the figure in Example 1.3, we note that there could possibly be different planes of orientation of the atoms in the crystal and the corresponding distances would be altered. For example, the planes could be chosen to pass through the atoms that are at the opposite corners of the cube rather than the horizontal or vertical planes treated here.

To interpret the measured Davisson–Germer signal, we will invoke some concepts of waves and particles and obtain the de Broglie wavelength. The group velocity of a wave as defined in the first section is $v_g = \partial\omega/\partial k$. Therefore, we write

$$v_g = \frac{\partial\omega}{\partial k} = \frac{\partial[(2\pi E)/h]}{\partial k} = \frac{1}{\hbar}\frac{\partial E}{\partial k}$$

where E is the energy of the wave and Equation (12) has been used. The velocity of a particle is computed from its energy $\mathscr{E} = p^2/2m$ as

$$\frac{\partial\mathscr{E}}{\partial p} = \frac{2p}{2m} = v_{\text{particle}}$$

Equate the group velocity of the wave with the particle velocity and let the energy of the wave E equal the energy of the particle $\mathscr{E}$. Therefore,

$$\frac{1}{\hbar}\frac{\partial E}{\partial k} = \frac{\partial \mathscr{E}}{\partial p} = \frac{\partial E}{\partial p}$$

whose integral is

$$p = \hbar k = \frac{h}{2\pi}\frac{2\pi}{\lambda}$$

This wavelength

$$\lambda_{\text{de Broglie}} = \frac{h}{p} \tag{16}$$

is called the de Broglie wavelength. This result is very profound in that de Broglie proposed that there is a duality in nature between waves and particles that becomes evident whenever the magnitude of h cannot be neglected. It was found that the experimental scattering of the particles from the crystal satisfied the relation

$$2d' \sin \phi_{\text{DG}} = n\lambda_{\text{de Broglie}} \tag{17}$$

where the velocity is taken to be the velocity of the electrons from the electron gun and n is an integer.

These two experiments highlight the development of thought that led to the wave–particle duality and hence to quantum mechanics. This led to the derivation of the Schrödinger equation.

III. Schrödinger Equation

From the experiments cited in Section II, we note the duality between waves and particles. It appears that we can use the mathematics garnered from one facet of nature to describe a feature found in another. Therefore, it would seem natural that we should be able to extend these descriptive techniques to a predictive mode in that something can now be stated about the properties of materials that will be later described. This in essence is what quantum mechanics will allow us to do. To derive the Schrödinger wave equation, we make use of some of the fundamentals of classical physics and the results of these experiments and their interpretation, which demonstrate this duality.

The total energy of a particle is

$$\mathscr{E} = \frac{p^2}{2m} + U$$

where the first term is its kinetic energy and the second term is its potential energy. The energy of a **photon** of light is

$$E = h\nu = \hbar\omega$$

Let us use the *ansatz* that these two energies are equal; that is,

$$E = \mathcal{E}$$

$$\hbar\omega = \frac{(h/\lambda_{\text{de Broglie}})^2}{2m} + U$$

where (16) has been employed to describe the momentum p. This can be written as

$$\hbar\omega = \frac{h^2 k^2}{2m} + U \tag{18}$$

where we freely interchange wave and particle effects and let $k = 2\pi/\lambda_{\text{de Broglie}}$.
    Let a wave be defined by

$$\psi = \psi_0 e^{i(kx - \omega t)} \tag{19}$$

The derivatives of (19) with respect to time and space can be written as

$$\frac{\partial \psi}{\partial t} = -i\omega\psi$$

$$\frac{\partial \psi}{\partial x} = ik\psi \tag{20}$$

$$\frac{\partial^2 \psi}{\partial x^2} = (ik)^2 \psi = -k^2 \psi$$

Multiply both sides of (18) by the wave function $\psi(x, t)$ given in (19). Hence we obtain

$$\hbar\omega\psi = \left(\frac{h^2 k^2}{2m}\right)\psi + U\psi \tag{21}$$

The crucial step now is to consider that the terms multiplying the wave function are actually *operating* on it. Therefore, from (21),

$$\frac{h}{-i}\frac{\partial \psi}{\partial t} = \frac{h^2}{2m}(-1)\frac{\partial^2 \psi}{\partial x^2} + U\psi$$

or

$$-i\hbar \frac{\partial \psi}{\partial t} = \frac{\hbar^2}{2m} \frac{\partial^2 \psi}{\partial x^2} - U\psi \tag{22}$$

This is the one-dimensional Schrödinger equation. In three dimensions, the spatial derivative term $\partial^2/\partial x^2$ is replaced with the Laplacian operator ∇^2, and we finally write

$$-i\hbar \frac{\partial \psi}{\partial t} = \frac{\hbar^2}{2m} \nabla^2 \psi - U\psi \tag{23}$$

This is the three-dimensional Schrödinger equation. Two problems remain. First, how does one obtain solutions to this partial differential equation and, second, what do solutions of this equation mean? The second question will be answered first, since solutions to an equation with no meaning will be of no value.

The wave function $\psi(x, t)$ does not have a meaning in itself. The function

$$\psi(x, t)\psi^*(x, t)\, dx = |\psi(x, t)|^2 \, dx$$

does, however, have a meaning in that it gives the probability that an object is located between x and $x + dx$. The star indicates the complex conjugate. The function $\psi(x, t)$ is normalized, so the probability is unity that the object is somewhere in the universe; thus

$$\int_{-\infty}^{\infty} \psi(x, t)\psi^*(x, t)\, dx = 1 \tag{24}$$

In three dimensions, this is written as

$$\int_{-\infty}^{\infty} \int_{-\infty}^{\infty} \int_{-\infty}^{\infty} \psi(\mathbf{r}, t)\psi^*(\mathbf{r}, t)\, dx\, dy\, dz = 1 \tag{25}$$

Example 1.4 Let the solution of the Schrödinger equation for a particular potential well be $\psi = e^{-|x/a|}e^{ivt}$. Calculate the normalized probability of finding a state between $a \leq x \leq 2a$.

Answer: The probability of finding a state between $a \leq x \leq 2a$ is defined as

$$\frac{\int_a^{2a} \psi\psi^*\, dx}{\int_{-\infty}^{\infty} \psi\psi^*\, dx} = \frac{\int_a^{2a} e^{-(2x/a)}\, dx}{2\int_0^{\infty} e^{-(2x/a)}\, dx}$$

The normalization integrand in the denominator is an even function of x, and we can remove the absolute value signs and integrate over one-half of the interval. The integrals can be performed, and we obtain the normalized probability:

$$\frac{-a/2(e^{-4} - e^{-2})}{2(-a/2)(0-1)} = \frac{e^{-2} - e^{-4}}{2} \approx 0.06$$

Quantum theory has some interesting predictions. For example, a car that is in very slow motion, sufficiently slow that it does not have enough energy to climb a hill, has, according to quantum theory, a nonzero probability of "tunneling" through the hill and appearing on the other side of it. This probability may be small but it is nonzero! As we will see later, a solid-state device, the tunnel diode, is based on this effect.

Solutions of the one-dimensional Schrödinger equation (22) can be obtained using the technique of separation of variables. In this technique, we assume that the wave function $\psi(x,t)$ can be expressed as

$$\psi(x,t) = X(x)T(t) \tag{26}$$

Substitute (26) in (22) and write

$$-i\hbar X(x)\frac{dT(t)}{dt} = \frac{\hbar^2}{2m}T(t)\frac{d^2 X(x)}{dx^2} - UX(x)T(t)$$

Divide both sides of this equation by $X(x)T(t)$ and write

$$-i\hbar\left(\frac{1}{T(t)}\right)\frac{dT(t)}{dt} = \frac{\hbar^2}{2m}\left(\frac{1}{X(x)}\right)\frac{d^2 X(x)}{dx^2} - U \tag{27}$$

where we have assumed that the potential U is only a function of position x. The left side of (27) is **independent** of x and the right side is **independent** of t. Hence each side can be replaced by a constant, which will be taken as $-E$. Therefore, (27) becomes two equations:

$$-i\hbar\frac{dT(t)}{dt} = -ET(t) \tag{28}$$

and

$$\frac{\hbar^2}{2m}\frac{d^2 X(x)}{dx^2} + (E - U)X(x) = 0 \tag{29}$$

Each of these equations can be solved separately. There is an important caveat that has been employed in writing (27). In dividing both sides of the equation by $X(x)T(t)$, an assumption has been made that neither of the terms will ever be zero; remember those warnings from the mathematicians. A mathematician, by inspection, would immediately write the separated equations and just state that (22) separates into two ordinary differential equations.

The solution of (28) is given by

$$T(t) = T_0 e^{i(E/\hbar)t} \tag{30}$$

where the frequency is $\omega = E/\hbar$. The solution of (29) can also be written as (this is valid for the potential energy $U = $ constant)

$$X(x) = A \sin kx + B \cos kx \tag{31}$$

where k is defined from

$$E - U = \frac{p^2}{2m} = \frac{(\hbar k)^2}{2m(2\pi)^2} \tag{32}$$

The constants of integration are specified from the boundary conditions of the potential well $(E - U)$ as a function of position x.

A particular potential well is shown in Figure 1.9, which is called an infinite well. No wave function can appear outside this well. The applicable boundary conditions for this well are

$$X(x = 0) = 0$$

$$X(x = L) = 0$$

$$X(x < 0) = X(x > L) = 0$$

The first of these boundary conditions dictates that the constant of integration B in (31) is equal to zero. The second specifies that

$$k = \frac{n\pi}{L}$$

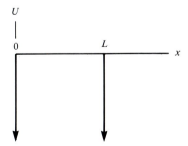

Figure 1.9   Infinite potential well.

where n is an integer. Therefore, from (32)

$$E - U = \frac{[h(n\pi/L)]^2}{2m(2\pi)^2} = \frac{n^2 h^2}{8mL^2} \tag{33}$$

Hence one notes that only **discrete** values of energy will satisfy this relation. The constant A can also be specified by the normalization that

$$\int_0^L |\psi|^2 \, dx = \int_0^L A^2 \sin^2\left(\frac{n\pi x}{L}\right) dx = 1$$

from which one finds that

$$A = \sqrt{\frac{2}{L}}$$

The solution of the one-dimensional Schrödinger wave equation (22) subject to an infinite potential well is finally given by

$$\psi(x, t) = X(x)T(t) = \sqrt{\frac{2}{L}} \sin\left(\frac{n\pi x}{L}\right) e^{i(E/\hbar)t} \tag{34}$$

Example 1.5 Solve the Schrödinger equation for an infinite potential well that has the boundary conditions that the normal derivative of the wave function is equal to zero at $x = 0$ and at $x = L$.

Answer: The general solution is given by the product of (30) and (31). Equation (30) describes the spatial variation from which we compute

$$\frac{dX(x)}{dx} = kA \cos kx - kB \sin kx$$

Since it is required that the normal derivative be zero at $x = 0$, the constant $A = 0$. Also, the normal derivative will be equal to zero if $k = n\pi/L$, where $n = 1, 2, 3, \ldots$. The final solution is

$$X(x) = \sqrt{\frac{2}{L}} \cos\frac{n\pi x}{L}$$

where the wave function has been normalized to 1.

The next step is to consider a one-dimensional potential well that is not inifinite in depth; that is, the wave function in the regions $x < 0$ and $x > L$ is not necessarily zero. In this case, as shown in Figure 1.10(a), there are three separate regions to be considered. The one-dimensional Schrödinger wave equation (22) has to be solved in each of the three regions separately. In doing this, we obtain

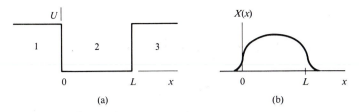

Figure 1.10   (a) A finite-depth potential well (b) Wave function for a finite-depth well.

six arbitrary constants of integration. The boundary conditions that can be immediately specified are that the solution should be continuous at $x = 0$ and at $x = L$. Also, we would expect that the solution should approach zero as x approaches ∞ and $-\infty$. It is reasonable to assume that the object should be located within or near the well. These are only four boundary conditions. Two more boundary conditions can be imposed from the assertion that the wave function solutions smoothly join each other at $x = 0$ and at $x = L$. This manifests itself with the requirement that $\partial X(x)/\partial x$ be continuous at $x = 0$ and at $x = L$. Note that the product solution (26) has been imposed and the temporal variation is separately treated.

The solution for $X(x)$ for the various regions can be written as

$$X(0 < x < L) = A \sin k_2 x + B \cos k_2 x$$
$$X(x < 0) = Ce^{k_1 x} + De^{-k_1 x} \tag{35}$$
$$X(x > L) = Ee^{k_3 x} + Fe^{-k_3 x}$$

The constants D and E are zero since the wave function is zero at $x = -\infty$ and at $x = \infty$, respectively. At $x = 0$, we obtain

$$B = C$$
$$k_2 A = k_1 C$$

At $x = L$, we obtain

$$A \sin k_2 L + B \cos k_2 L = Fe^{-k_3 L}$$
$$k_2 A \cos k_2 L - k_2 B \sin k_2 L = -k_3 Fe^{-k_3 L}$$

In both of these sets of expressions, the continuity of the function and its derivative have been imposed. The normalization given in (24) will still be imposed. A sketch of the resulting wave functions is shown in Figure 1.10(b).

It is possible to repeat this calculation in three dimensions in rectangular coordinates in a straightforward fashion. The reader will wonder if it is natural to assume that atoms are cubical in shape. The answer is **no**, but the mathematics is much easier. We assume that the three-dimensional wave equation (23) separates with the assumption

$$\psi(x, y, z, t) = X(x)Y(y)Z(z)T(t)$$

and obtain four ordinary differential equations. Using separation of variables with a cubical volume of $L \times L \times L$ with the boundary condition that the wave function is zero on all surfaces, we compute the normalized wave function to be

$$\psi(x, y, z, t) = \sqrt{\frac{8}{L^3}} \sin\frac{n_x \pi x}{L} \sin\frac{n_y \pi y}{L} \sin\frac{n_z \pi z}{L} e^{-i(E/\hbar)t}$$

where

$$E - U = \left(\frac{h^2}{8mL}\right)^2 \left[\left(\frac{n_x \pi}{L}\right)^2 + \left(\frac{n_y \pi}{L}\right)^2 + \left(\frac{n_z \pi}{L}\right)^2\right] \tag{36}$$

In addition to these three numbers, an additional number, n_s, reflects the fact that there is an ambiguity in which direction an electron spins (i.e., clockwise or counterclockwise). Therefore, we write that $n_s = +(\frac{1}{2})$ and $n_s = -(\frac{1}{2})$. The numbers n_x, n_y, n_z, and n_s are called the **quantum numbers**. Pauli argued that no two electrons can have the same two quantum numbers (exclusion principle).

But an atom is generally accepted to have a spherical shape since the electrons can be "compacted" into the smallest volume. Hence the three-dimensional wave equation (23) should be solved in spherical coordinates rather than rectangular coordinates as we have done. The procedure is straightforward and the details are usually treated in advanced courses in mathematical methods or quantum mechanics. The Schrödinger equation separates into four equations, one that depends on time and one each that depend on radius and the two angular coordinates individually. We turn to the equations that were studied by Bessel and Legendre and find that there are certain prescribed relations between the separation constants or the quantum numbers. (We will employ the commonly used notation for stating them.) Without presenting the details, these relations can be written as

$$\begin{aligned}
&\mathbf{n} = 1, 2, 3, \ldots \\
&\mathbf{l} = 0, 1, 2, \ldots, (\mathbf{n} - 1) \\
&\mathbf{m} = -1, \ldots, -2, -1, 0, 1, 2, \ldots, +1 \\
&\mathbf{n_s} = +\tfrac{1}{2} \quad \text{and} \quad -\tfrac{1}{2}
\end{aligned} \tag{37}$$

where the last number reflects the ambiguity of the spin's being clockwise or counterclockwise. We have now found the quantum numbers and hence the allowed states for the electrons in an atom. The distribution of these electrons can be shown with the following table (Table 1.1) for values of $\mathbf{n}$ up to 3.

TABLE 1.1

n	l	m	n_s	Allowable States in Subshell	Allowable States in Completed Shell
1	0	0	$\pm\frac{1}{2}$	2	2
2	0	0	$\pm\frac{1}{2}$	2	
					8
	1	−1	$\pm\frac{1}{2}$		
		0	$\pm\frac{1}{2}$	6	
		+1	$\pm\frac{1}{2}$		
3	0	0	$\frac{1}{2}$	2	
	1	−1	$\pm\frac{1}{2}$		
		0	$\pm\frac{1}{2}$	6	
		+1	$\pm\frac{1}{2}$		
					18
	2	−2	$\pm\frac{1}{2}$		
		−1	$\pm\frac{1}{2}$		
		0	$\pm\frac{1}{2}$	10	
		+1	$\pm\frac{1}{2}$		
		+2	$\pm\frac{1}{2}$		

From this state table for the hydrogen atom, we note that the electron can occupy several states in addition to the lowest (*ground*) state. In fact, we are now prepared to classify atoms according to the configuration of their electrons in the ground state in Table 1.2.

In the work described, we were certain about our predictions concerning a probability distribution function and hence were certain about the location and momentum of our states and therefore the electrons. This unfortunately is not true, and a certain degree of ambiguity enters in, which is called the Heisenberg uncertainty principle. Several expressions can be written to express it, and two are

$$\Delta p \, \Delta x > h$$

$$\Delta E \, \Delta t > h$$

They state that the uncertainty in knowing the exact values of two quantities that are related must be greater than Planck's constant. These related variables are sometimes called *conjugate variables*.

To illustrate the first of these relations, let us consider the situation of a ball slowing down from a velocity $(v + \Delta v)$ to a velocity v. As it slows down, it emits a photon. Therefore, from the conservation of energy,

$$\frac{m}{2}(v + \Delta v)^2 - \frac{m}{2}v^2 = hv$$

TABLE 1.2

Atomic Number (Z)	Element	n = 1, l = 0	n = 2, l = 0	n = 2, l = 1	n = 3, l = 0	n = 3, l = 1	n = 3, l = 2	n = 4, l = 0	n = 4, l = 1
		Number of Electrons							
1	H	1							
2	He	2							
3	Li		1						
4	Be		2						
5	B		2	1					
6	C	Filled	2	2					
7	N		2	3					
8	O		2	4					
9	F		2	5					
10	Ne		2	6					
11	Na				1				
12	Mg				2				
13	Al				2	1			
14	Si				2	2			
15	P	Filled	Filled		2	3			
16	S				2	4			
17	Cl				2	5			
18	Ar				2	6			
19	K							1	
20	Ca							2	
21	Sc						1	2	
22	Ti						2	2	
23	V						3	2	
24	Cr						4	2	
25	Mn						5	2	
26	Fe						6	2	
27	Co	Filled	Filled		Filled		7	2	
28	Ni						8	2	
29	Cu						10	1	
30	Zn						10	2	
31	Ga						10	2	1
32	Ge						10	2	2
33	As						10	2	3
34	Se						10	2	4
35	Br						10	2	5
36	Kr						10	2	6

This can be approximated as

$$mv\,\Delta v = hv \approx h\frac{v}{\lambda}$$

or

$$\lambda m\,\Delta v \approx h$$

But to measure the location of the ball, we need a measuring stick whose finest gradation is less than the expected deviation in position. For example, the measurement of the size of a pin cannot be accomplished with any accuracy with a meter stick. Let us assume that the deviation of position $\Delta x > \lambda$, where λ is the wavelength of the diagnosing wave. Finally, let $m\,\Delta v = \Delta p$ and we obtain

$$\Delta x\,\Delta p > \lambda\,\Delta p \approx h$$

This calculation is of the "back-of-the-envelope" type, but it does illustrate the Heisenberg uncertainty principle.

IV. Conclusion

Quantum mechanics has given us a new way of thinking in that it conjoins the two separate fields of wave physics and particle physics and leads to the predictive capability of explaining the atom. The highlights of quantum mechanics outlined here form the basis of solid-state physics, which allows one to predict the possible states or locations where electrons can reside in the region about the central nucleus. Whether an electron actually fills these states is a subject that will be examined in the next chapters.

PROBLEMS

1. Show that a solution of the wave equation can be given by:
   (a) $\psi = \psi_0 \sin(kx - \omega t)$.
   (b) $\psi = \psi_0 e^{-(k|x-ct|)}$; define c.
2. Sketch the wave solutions given in Problem 1 at three values of time if $c = 2$.
3. Consider a distributed transmission line consisting of sections made of resistors and capacitors as shown in the following sketch:

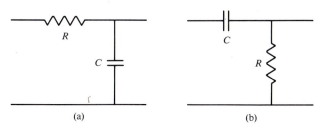

(a)   (b)

(a) Derive the partial differential equations that describe each transmission line.

(b) Assume that a sinusoidal signal $\phi = \phi_0 e^{i\omega t}$ is connected at one end of the transmission line. Find the dispersion relation for the two lines.

(c) Calculate the phase and group velocities for each line.

(d) Assume that a detecting light bulb that turns on if the voltage is greater than ϕ_a is connected across the shunt element. Let a battery be connected through a switch at one end of either of the lines. If the switch is closed at $t = 0$, interpret the resulting wave of light or darkness in terms of the phase and group velocities. You may wish to perform this experiment.

4. The Laplacian operator ∇^2 in spherical coordinates is given by

$$\nabla^2 f = \frac{1}{r^2}\frac{\partial}{\partial r}\left(r^2 \frac{\partial f}{\partial r}\right) + \frac{1}{r^2 \sin\psi}\frac{\partial}{\partial\psi}\left(\sin\psi \frac{\partial f}{\partial\psi}\right) + \frac{1}{r^2 \sin\psi}\frac{\partial^2 f}{\partial\theta^2}$$

(a) Show that the wave equation separates into three ordinary differential equations.

(b) Find the solution for f if there is no θ or ϕ dependence. Interpret the solution.

5. Show that an electron that is brought from ∞ to a distance r from a positive charge has a potential energy given by $q^2/4\pi\varepsilon_0 r$ if the charges have an equal magnitude.

6. Calculate the values of the first three discrete energy differences between adjacent levels given by equation (14).

7. For the KCl crystal described in Example 1.3, calculate the distance between diagonal parallel planes that pass through the atoms at the opposite corners of the cube.

8. Calculate the distance between the adjacent atoms of the following crystal. The weight of each atom is W. The linear dimensions of each small cube are $d' \times d' \times d'$.

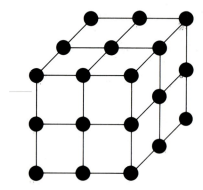

9. It is determined that the dispersion relation for a certain media is $\omega = ak - bk^3$. Find the partial differential equation that yielded this dispersion relation.

10. A dispersion relation is given by $\omega = \omega_0 \sin k/k_0$. Sketch this relation and, using approximations, find the describing partial differential equation. Also sketch the approximations.

11. Solve the one-dimensional Schrödinger equation for the potential well indicated in the following figure. The potential U is $-\infty$ at $x = 0$ and at $x = L$.

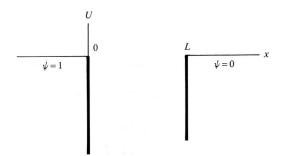

12. A solution of the Schrödinger equation is found to be

$$\psi = \psi_0 \sin\left(\frac{n\pi x}{L}\right) e^{ivt} \qquad (0 \leq x \leq L)$$

$\psi = 0$ elsewhere. Calculate the normalized probability of finding a state in the region $0 \leq x \leq L/2$ for various values of n.

13. Derive the solution for a finite-depth potential well as shown in Figure 1.9 and accurately sketch the solution for $X(x)$.

14. Derive equation (36) for a cubical volume potential well.

15. The potential structure surrounding a hydrogen molecule can be written as

$$U(x) = -\frac{q^2}{4\pi\varepsilon_0}\left[\frac{1}{|x - x_0|} + \frac{1}{|x + x_0|}\right]$$

where $2x_0$ is the separation of the two hydrogen nucleii.
    (a) Sketch this potential variation.
    (b) Sketch the expected variation of the magnitude of the solution of the Schrödinger equation for this potential.
    (c) Discuss the effect of having two potential wells in close juxtaposition.

Chapter Two

Statistical Quantities

The collision of two balls in space is a problem that can be solved easily by the most casual of observers, and we can write down the position, the momentum, and so on, even if we incorporate the gravitational forces that exist between the balls. However, there is as yet no known general analytical solution for three bodies that have collided under the condition that there is some force among them. Oh yes, there are numerical solutions, but nothing is known analytically. Imagine a dust particle floating around Saturn being under the influence of both Saturn and its moon. It moves in a random but prescribed path. Add more dust particles and we obtain a ring structure that gets more complicated as the details are brought out through further satellite investigations. The word *stochastic* has entered the vocabulary of people who investigate these problems. Since even the three-body problem has no solution, what hope is there in explaining the inner workings of a solid-state device? There is a more than a modicum of hope, since the reader can use the tools of statistics and obtain a reasonable prediction of what to expect under certain well-prescribed conditions.

Certain constraints may be imposed on a system that will allow a solution to be obtained. The simplest of these is the constraint of *thermodynamic equilibrium*. This is fundamental and is associated with the concept that a state will remain as is unless it is acted on by an external force. An individual state may change, but a distribution of states will be stable.

The understanding of various statistical distributions will give the reader some confidence in the predictive capabilities that are needed in order to develop and design components that become smaller in size but more complex in capabilities. Three of these distributions will be developed here after some of the properties of statistics are illustrated.

I. Statistics

Before embarking on a long discourse on the details of thermodynamic distributions and expected effects, we will first review with an example some of the basic definitions of statistics. The reader has probably encountered the "flipping of a coin" in the journey through life and has decided that the chances of obtaining heads or tails are evenly divided. We could assert that we had a bimodal distribution, which is not too interesting as a thermodynamic descrip-

TABLE 2.1

					6–1					
				5–1	5–2	5–3				
			4–1	4–2	4–3	4–4	4–5			
		3–1	3–2	3–3	3–4	3–5	3–6	4–6		
	2–1	2–2	2–3	2–4	2–5	2–6	5–4	5–5	5–6	
1–1	1–2	1–3	1–4	1–5	1–6	6–2	6–3	6–4	6–5	6–6
2	**3**	**4**	**5**	**6**	**7**	**8**	**9**	**10**	**11**	**12**

tion. Let us examine a slightly more complicated example, that of throwing a pair of dice (we will assume that they are unbiased for the moment). In Table 2.1, the possible combinations are given.

The overall combination of numbers is called a *distribution*. It does not depend on the individual components in that they can be interchanged and it will not alter the overall complexion of the distribution. For example, 1–6 and 6–1 both yield an item in bin 7. The probability of obtaining a certain number could be found from

$$P = \frac{\text{number of possible combinations of a certain number}}{\text{total combinations}}$$

For example, the probability of obtaining a 5 is $\frac{4}{36}$. Let us imagine a game where we desire to throw a 5 twice in a row. The probability of this would be

$$\frac{4}{36} \times \frac{4}{36} = \frac{1}{81}$$

The fancy hotels in gambling cities and state lotteries support themselves by a deep understanding of the "odds."

This distribution has assumed that there was no bias in the statistics. Let us repeat it with a bias by adding a weight to one of the dice such that it is usually greater than 3. The distribution will then be distorted such that the bins that contain the numbers 2 and 3 will be less frequently populated than we would expect in an honest game. The statistics that we will study will be based on physical intuition.

II. Maxwell–Boltzmann Distribution

Nature has been kind to us in that it knows the ultimate fate of things; they will reach thermal equilibrium through irreversible processes. In this state, the entropy that is a measure of the random behavior of particle motion reaches a maximum value. The entropy of a macrostate S is related to the number of microstates W that lead to it through the relation

$$S = k_B \ln W \qquad (1)$$

where k_B is Boltzmann's constant ($k_B = 1.38 \times 10^{-23}$ joule/°K) and ln is the natural logarithm. Two consequences of thermal equilibrium are that any macroscopic quantity can have no spatial variation and there can be no ordered motion in the system. Hence no external forces can act on the system. A current flowing in a resistor would not be in thermal equilibrium.

Having introduced the concept of entropy at this stage, it is useful to recall two laws of thermodynamics that will be used later. The first states that energy must be conserved, and the second states that the entropy of a system will increase until an equilibrium situation occurs and the entropy is a maximum.

Why study the equilibrium state then since it appears to describe such a seemingly uninteresting environment? First, it is relatively simple, and, second, there are systems that are close to thermal equilibrium such that we may be able to carry forth a derivation that examines perturbations about the equilibrium value.

Let us first obtain an understanding of the definition of a microstate W. This can be effected with an example. Assume that one possesses three numbered balls. How many different ways are there to organize them?

1	2	3
2	3	1
3	1	2
1	3	2
2	1	3
3	2	1

There are six possible ways of organizing the three balls. We note that $3! = 6$. This could be generalized to N_j balls that are each individually identified with a number, so there are $N_j!$ different ways of organizing them in a bin whose name is j. The total number of bins is J.

Let us extend this to five balls and three pails. They could be distributed in the three pails as

a.	1–2	3	4–5
b.	2–1	3	4–5
c.	1–3	2	4–5

In cases a and b, each microstate (pail) has the same balls, although they are in a different order. The microstates are the same. In case c, the microstates are different. However, the same five balls have been used in each case so the macrostates are the same. In general, each microstate j can have $N_j!$ ways of organizing the numbered balls within it. The total number of balls is equal to N.

The total number of distinct microstates that can lead to the same macrostate is therefore

$$W = \frac{N!}{(N_1!)(N_2!)\ldots(N_j!)\ldots(N_J!)} \tag{2}$$

or

$$W = \frac{N!}{\prod\limits_{j=1}^{J}(N_j!)} \tag{3}$$

where $\prod$ indicates a repeated product and there are J volumes. Substitute (3) in (1):

$$S = k_B \ln \frac{N!}{\prod\limits_{j=1}^{J}(N_j!)} \tag{4}$$

This is the entropy.

Example 2.1   Calculate the entropy for a system containing five numbered balls distributed in three pails in which the first pail holds two balls, the second pail holds one ball, and the third holds two balls.

Answer: Since one can interchange the order of the balls without changing the microstates, we write that in the first pail there are two ways of reorganizing the balls, one in the second pail, and two in the third pail. Therefore,

$$W = \frac{N!}{(N_1!)(N_2!)(N_3!)} = \frac{5!}{(2!)(1!)(2!)} = 30$$

and $S = k_B \ln W = k_B \ln 30$.

The requirement for thermal equilibrium is that the entropy must be a maximum, but this requirement is subject to two constraints. The first constraint is that the total number of particles in the system be a constant:

$$\sum_{j=1}^{J} N_j = N = \text{constant} \tag{5}$$

The second constraint is that the energy of the system be a constant:

$$\sum_{j=1}^{J} E_j N_j = \mathscr{E} = \text{constant} \tag{6}$$

where $E_j = p_j^2/2m$. To find the maximum of a function subject to constraints, we use the technique of Lagrange multipliers.

Lagrange multipliers can be illustrated through an example. Calculate the dimensions of a rectangular garden that has the maximum area that can be enclosed within a fence whose length is fixed. Let the dimensions be a and b and the length of the fence be c. Therefore, $A = ab$ with the constraint $c = 2a + 2b$. The procedure of using Lagrange multipliers is to seek a path to the top of a hill that is defined by

$$\text{Hill} = \text{area} - \mathscr{L} \text{ (constraints)}$$

where $\mathscr{L}$ is the unknown Lagrange multiplier. This is shown in Figure 2.1. In terms of the example,

$$H = ab - \mathscr{L}(2a + 2b - c)$$

where H represents the hill. The highest point on the hill is determined from the simultaneous satisfaction of

$$\frac{\partial H}{\partial a} = 0 = b - 2\mathscr{L}$$

$$\frac{\partial H}{\partial b} = 0 = a - 2\mathscr{L}$$

From this, we compute that

$$a = b = \sqrt{A}$$

Therefore, the garden will be square in shape. The number of Lagrange

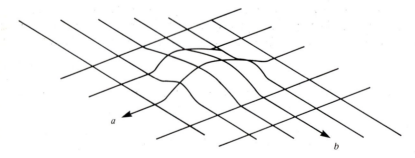

Figure 2.1 Example of the usage of Lagrange multipliers.

multipliers $\mathscr{L}_j$ equals the number of constraints imposed on the equation. For our thermodynamics problem, there are two constraints.

Therefore, the maximum of the entropy (4) subject to the constraints (5) and (6) is computed from

$$H = k_B \left[\ln N! - \ln \left(\prod_{j=1}^{J} (N_j!) \right) \right] - \mathscr{L}_1 \left(N - \sum_{j=1}^{J} N_j \right) - \mathscr{L}_2 \left(\mathscr{E} - \sum_{j=1}^{J} E_j N_j \right)$$

Stirling's formula allows us to approximate $\ln n!$ for large n as

$$\ln n! \approx n \ln n - n$$

and H can be written as

$$H \approx k_B \left\{ [N \ln N! - N] - \sum_{j=1}^{J} [N_j \ln N_j - N_j] \right\}$$

$$- \mathscr{L}_1 \left(N - \sum_{j=1}^{J} N_j \right) - \mathscr{L}_2 \left(\mathscr{E} - \sum_{j=1}^{J} E_j N_j \right) \tag{7}$$

We calculate the maximum from

$$\frac{\partial H}{\partial N_j} = 0 = \sum_{j=1}^{J} \left\{ -k_B \left[\ln N_j + \frac{N_j}{N_j} - 1 \right] + \mathscr{L}_1 + \mathscr{L}_2 E_j \right\}$$

or

$$0 = \sum_{j=1}^{J} \left\{ -k_B \ln N_j + \mathscr{L}_1 + \mathscr{L}_2 E_j \right\} \tag{8}$$

For this to be satisfied for all values of j, then $\{ -k_B \ln N_j + \mathscr{L}_1 + \mathscr{L}_2 E_j \}$ must be equal to zero. This can also be written as

$$N_j = \exp \left\{ \frac{\mathscr{L}_1 + \mathscr{L}_2 E_j}{k_B} \right\} \tag{9}$$

Equation (9) was derived considering discrete particles each with discrete energies. If we assume that a gas is being studied with a distribution of energies between E_i and $E_i + dE_i$, the discrete nature of (9) becomes a continuous distribution

$$f = \exp \left\{ \frac{\mathscr{L}_1 + \mathscr{L}_2 E}{k_B} \right\}$$

Let us now calculate the Lagrange multiplier $\mathscr{L}_2$. Equation (4) can be written as

$$
\begin{aligned}
S &= k_B \left\{ \ln N! - \ln \left[\prod_{j=1}^{J} N_j! \right] \right\} \\
&= k_B \left\{ \ln N! - \sum_{j=1}^{J} \ln N_j! \right\} \\
&\approx k_B \left\{ \ln N! - \sum_{j=1}^{J} (N_j \ln N_j - N_j) \right\}
\end{aligned}
$$

where Stirling's formula has been applied. Let us find the maximum value of this equation:

$$
\Delta S \approx k_B \left\{ \Delta(\ln N!) - \sum_{j=1}^{J} \left[\Delta N_j \ln N_j + N_j \left(\frac{\Delta N_j}{N_j} \right) - \Delta N_j \right] \right\}
$$

The first term is zero since N is a constant. Substitute (9) in this equation and write

$$
\Delta S \approx - \sum_{j=1}^{J} (\mathscr{L}_1 + \mathscr{L}_2 E_j) \Delta N_j = - \mathscr{L}_2 \sum_{j=1}^{J} E_j \Delta N_j
$$

since the summation of the derivative ΔN_j is equal to zero. This assumes that, on the average, the deviation about the equilibrium is zero. Therefore, the deviation of the entropy from its equilibrium value is

$$
\Delta S = - \mathscr{L}_2 \Delta E
$$

From the first law of thermodynamics, we write the conservation of energy as

$$
\Delta E = \Delta Q - \Delta W
$$

where ΔE is the internal energy of the gas, $\Delta Q = T \Delta S$ is defined as the heat into the gas, and $\Delta W = F \Delta x = (F/A) A \, \Delta x = P \, \Delta V$ is the work (P is pressure and ΔV is the change in volume). If the volume is constant, then

$$
\Delta S = \frac{\Delta E}{T}
$$

The second law of thermodynamics states that $\Delta Q \leq T \, \Delta S$. Therefore,

$$
\mathscr{L}_2 = - \frac{1}{T}
$$

The Lagrange multipliers are now identified as $e^{\mathcal{L}_1/k_B}$, which is a constant and $\mathcal{L}_2 = -1/T$, which is the average thermal energy per particle. Hence

$$f = e^{\mathcal{L}_1/k_B} e^{-E/k_B T} \tag{10}$$

which is almost the Maxwell–Boltzmann distribution, where the kinetic energy is $E = mv^2/2$ and the thermal velocity is defined from $v_{\text{th}}^2 = k_B T/m$.

To evaluate the constant term $e^{\mathcal{L}_1/k_B}$, we make use of the definition of the total number of particles:

$$N = \sum_{j=1}^{J} N_j$$

$$= \sum_{j=1}^{J} e^{\mathcal{L}_1/k_B} e^{-E_j/k_B T}$$

From this, we write

$$e^{\mathcal{L}_1/k_B} = \frac{N}{Z}$$

where Z is called the partition function and

$$Z = \sum_{j=1}^{\infty} e^{(-p_j^2/2mk_B T)} \rightarrow \left(\frac{1}{\mu}\right) \int\int e^{(-p^2/2mk_B T)} \, d\mathbf{r} \, d\mathbf{p}$$

where in the continuum limit the summation becomes an integral over a six-dimensional phase space and μ is an incremental volume such that the phase space density $f_{\text{MB}} = N_j/\mu$. The integral can be performed to yield

$$f_{\text{MB}} = \frac{N/V}{(2\pi mk_B T)^{3/2}} e^{(-p^2/2mk_B T)} \tag{11}$$

Equation (11) is the Maxwell–Boltzmann distribution that gives the equilibrium distribution of particles. It is shown in Figure 2.2.

Two quantities can be immediately calculated. The first is the number of particles in the system, which is calculated from

$$N = \int_{-\infty}^{\infty} f_{\text{MB}} \, dp$$

The second is the current density of these particles:

$$j = q \int_{-\infty}^{\infty} p f_{\text{MB}} \, dp$$

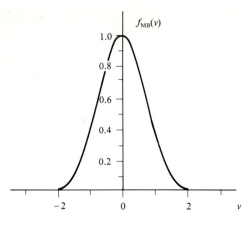

Figure 2.2   Maxwell–Boltzmann distribution.

where q is the charge. The current is equal to zero, first by the definition of equilibrium thermodynamics, since no currents can exist, and second by the following mathematical argument. The momentum p is an odd function in the variable p and the Maxwell–Boltzmann distribution is an even function. Hence the product of an even function and an odd function is an odd function, and the integral over all space of an odd function (or over a period of a periodic function) is zero. For example, the function $\sin x$ is an odd function and the integral of it from $-x_0$ to $+x_0$, where x_0 could be $\pi/2$, is zero.

Example 2.2   Show that the current carried in a device in thermal equilibrium is identically zero.

Answer: The current density j is defined as $j = q \int_{-\infty}^{\infty} pf \, dp$, where f is the distribution function of the particles. Since it is in thermal equilibrium, we write $f = f_{MB} = A \exp[-Bp^2]$, where A and B are constants. Therefore, we sketch the product which appears in the integrand as shown in the following figure:

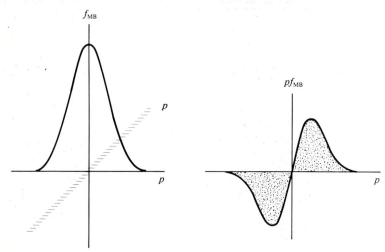

Note that the area under the curve $p > 0$ equals that under the curve $p < 0$ but has the opposite sign. Therefore, the net area is zero and the current that is proportional to the net area is also zero.

Therefore, from first principles, we have derived the distribution of particles in thermal equilibrium to be the Maxwell–Boltzmann distribution given in Equation (11). In the derivation, we have made the assumption that the individual particles can be identified and have a character all their own. If this assumption of identifiable particles is removed, the Maxwell–Boltzmann distribution function is modified. This modification was carried out by Bose and Einstein. Following the procedure outlined previously, they showed that the distribution (11) is modified such that the exponential term is changed as

$$\exp\left(\frac{-p^2}{2mk_BT}\right) \rightarrow \frac{1}{\exp((p^2/2m) - (E_B/k_BT)) - 1}$$

where E_B is the Bose energy. In the limit of large energies, this distribution reduces to the Maxwell–Boltzmann distribution.

Neither of these distributions has made any restriction on the number of particles that can occupy a given energy state. Therefore, neither distribution has taken into account the Pauli exclusion principle that limits this number to 1. This was accomplished by Fermi and Dirac.

III. Fermi–Dirac Distribution

The Fermi–Dirac distribution is the most important of the distributions for the understanding of solid-state physics. In agreement with the Maxwell–Boltzmann statistics and the Bose–Einstein statistics, each state is also assumed to be statistically independent (i.e., noninteracting). Also, each particle is not uniquely identified as is also found in Bose–Einstein statistics. However, in contrast to the previous distributions, Fermi and Dirac incorporated the statement that only one particle could occupy any given energy state. This distribution will be encountered frequently throughout the material in this book.

The procedure used to derive the Fermi–Dirac distribution is similar to the procedure that was used to derive the Maxwell–Boltzmann distribution in that we will again maximize the resulting entropy, but will include the effects of the Pauli exclusion principle, which states that no two particles can have the same state. The phase space will be divided into volumes $d\mathbf{r}\,d\mathbf{p}$ such that all states within the volume have the same energy. The number of particles in this six-dimensional phase space volume will be a macroscopic variable. In each of these volumes, there will be G_j states that are occupied by N_j electrons. Obviously, $N_j \le G_j$. The equilibrium value of N_j will now be computed.

The number of ways that N_j particles can be distributed among the G_j states can be written as

$$G_j(G_j - 1)(G_j - 2)\ldots(G_j - N_j + 1) = \frac{G_j!}{(G_j - N_j)!}$$

These particles all have the same energy E_j. This equation can be interpreted in the following manner. The first particle can reside in G_j states, the second in the $G_j - 1$ states that remain, the third in the $G_j - 2$ states that remain, and so on. However, the particles cannot be distinguished from each other, so many of these permutations lead to the same wave function. Therefore, the number of ways that N_j particles can reside in G_j states is

$$W_j = \frac{G_j!}{(N_j!)(G_j - N_j)!} \tag{12}$$

N indistinguishable electrons can be arranged in J volumes, each with a different energy as

$$W = \prod_{j=1}^{J} W_j \tag{13}$$

Hence, the entropy is computed from equation (1) as

$$S = k_B \sum_{j=1}^{J} \{\ln G_j! - \ln N_j! - \ln(G_j - N_j)!\} \tag{14}$$

The electrons still must satisfy the constraints concerning the conservation of particles (5) and energy (6) that were employed in deriving the Maxwell–Boltzmann distribution. The equivalent equation to (7) with the unknown Lagrange multipliers is written as

$$H \approx k_B \sum_{j=1}^{J} \{[G_j \ln G_j - G_j] - [N_j \ln N_j - N_j] - [(G_j - N_j)\ln(G_j - N_j) - (G_j - N_j)]\}$$

$$- \mathscr{L}_1\left(N - \sum_{j=1}^{J} N_j\right) - \mathscr{L}_2\left(\mathscr{E} - \sum_{j=1}^{J} E_j N_j\right)$$

where Stirling's formula has again been applied. The maximum of H is found from

$$\frac{\partial H}{\partial N_j} = 0 = \sum_{j=1}^{J} \{-k_B \ln N_j + k_B \ln(G_j - N_j) + \mathscr{L}_1 + \mathscr{L}_2 E_j\}$$

For this to be valid for any value of j, the term within the summation

$$-k_B \ln N_j + k_B \ln(G_j - N_j) + \mathscr{L}_1 + \mathscr{L}_2 E_j$$

must equal zero. This can be written as

$$\ln\left\{\frac{G_j - N_j}{N_j}\right\} = \frac{-1}{k_B}(\mathscr{L}_1 + \mathscr{L}_2 E_j)$$

Solve for N_j and write

$$N_j = \frac{G_j}{1 + \exp\{(-1/k_B)(\mathscr{L}_1 + \mathscr{L}_2 E_j)\}} \tag{15}$$

The probability of finding a state occupied is therefore equal to the number of particles N_j divided by the number of states G_j.

The Lagrange multipliers are determined by following the procedure detailed in the previous section. We find that $\mathscr{L}_2 = -1/T$. It is common to define $\mathscr{L}_1$ as $\mathscr{L}_1 = E_F/T$, where E_F is known as the Fermi energy. These two Lagrange multipliers are incorporated into (15), and we find the probability that a state actually contains an electron is given by

$$f(E) = \frac{1}{1 + \exp\{(E - E_F)/k_B T\}} \tag{16}$$

where $E = p^2/2m$. This function is called the Fermi–Dirac function. It is shown in Figure 2.3 for various values of temperature T. At large values of energy, $E \gg E_F$, $f(E)$ reduces to the Maxwell–Boltzmann distribution.

The Fermi energy in thermal equilibrium is equal to the chemical potential. The chemical potential is the amount of energy by which the net energy of the system changes as particles are added to or subtracted from it.

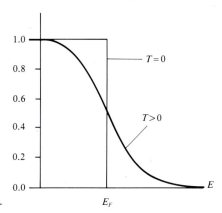

Figure 2.3   Fermi–Dirac distribution.

IV. Conclusion

Thermodynamics has given us a statistical description for the distribution of noninteracting particles in an ideal gas. A description was formulated for both classical and quantum systems. For a classical system, the particles are assumed to be distinguishable, and any number of particles can exist in the same state. The resulting distribution is the Maxwell–Boltzmann distribution. If we start with another distribution, it will evolve into this Maxwell–Boltzmann distribution. The process for this evolution can be described with a kinetic equation, which will be described in the next chapter.

For a quantum system, the particles are assumed to be indistinguishable. In the case of any number of particles existing in a given state, we obtained the Bose–Einstein distribution. The quatum system that incorporates the additional requirement that no more than one particle can exist in a state led to the Fermi–Dirac distribution.

Therefore, quantum mechanics has given us a prescription for finding the states where an electron could exist in an atom. Each of these states has a unique identity that can be specified by its quantum numbers. Thermodynamics has told us the probability that electrons actually do reside·in these states. The restriction that will be imposed on the system is that no more than one electron can occupy the same state. The next question that must be raised is how these atoms and electrons are distributed in a material. This will be examined in detail later when the band structure of materials will be detailed and the various types of materials will be distinguished.

PROBLEMS

1. Using Lagrange multipliers, show that a rectangular box whose dimensions are $a \times b \times c$ and whose surface area is a constant will have the maximum volume if the box has the form of a cube.

2. Stirling's formula $\ln n! \approx n \ln n - n$ allowed us to obtained the Maxwell–Boltzmann distribution. Estimate the error in this formula for $n = 1; 10; 100; 1000$.

3. A distribution function that is simpler to use in many applications is the "water-bag" distribution shown in the following figure:

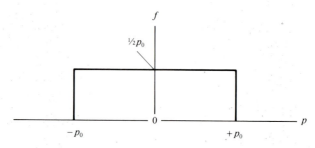

(a) Calculate the density of particles associated with this distribution.

(b) Calculate the current associated with this distribution.

(c) Let this rectangular box now be centered about the momentum $+p_0$ and repeat parts (a) and (b). Is this in equilibrium? Why?

4. Repeat Problem 3 if the distribution has a triangular shape.

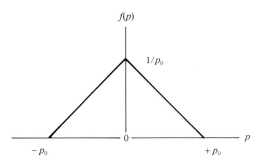

5. By carrying out the integrations, show that the current carried in a media in thermal equilibrium is equal to zero.

6. Calculate the number of particles that reside in a state whose energy is greater than the Fermi energy if the temperature of the system is $0°K$.

7. It is found that states are distributed in a material with an energy distribution $g = g_0 \exp[(E - E_c)/k_B T]$ for $E \geq E_c$ and $g = 0$ for $E < E_c$. Calculate the number of electrons occupying these states at $0°K$ if (a) $E_c < E_F$, and (b) $E_c > E_F$. Sketch the distribution function and the Fermi function for each case.

Chapter Three

Nonequilibrium Statistics

In the previous chapter, the long time evolution of natural phenomena was shown to be described by a Maxwellian distribution f_{MB}. The reader probably thought that this was rather heavy mathematics and, after a long calculation, obtained a distribution that was valid only in a limit in time that most mortals will never see. Yes, one could fret about this equilibrium value, but this may not be the troubling point. The more serious question is, "How does today's distribution that may exist in some local region in space evolve into this equilibrium distribution that cannot depend on space?"

The distribution will evolve through some irreversible process before it reaches the equilibrium value. A simple example of such a process would be passing a current through a resistor and noting the radiation of heat. The opposite procedure of putting a resistor in an oven will not generate a current. Another irreversible process is the aging process; no one is getting younger despite the age-old attempts of people. The vanity business lives on these attempts.

To study this temporal evolution of the initial distribution function into a Maxwellian distribution, we will first examine the phenomena of the collision of particles and then the process of diffusion of a group of particles. This will lead to the kinetic equation that will allow us to predict the behavior of the distribution as it changes from an initial condition to the asymptotic state of the Maxwell–Boltzmann distribution.

I. Collisions

As a particle meanders in a gas, there is a chance that it will encounter another like or unlike particle and possibly interact with it. This collision could manifest itself in several possible scenarios, from two particles "just passing in the wind" and not interacting, to a violent interaction such that the particles could exchange a charge or one particle could ionize the other. There are certain terms that are common to all types of collisions and they will be obtained here.

The first term is the **cross section σ_c** of the collision. Imagine that two billiard balls pass near each other as shown in Figure 3.1. Let us neglect the gravitational attraction between the balls. In this case, the balls will interact only if they pass within a dimension equal to the sum of the radii of the balls. The cross section of the collision is equal to $\pi(2R)^2$ if the balls have the same radius. This is greater

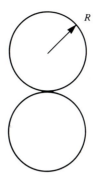

Figure 3.1 Collision of two balls whose individual radius is *R*.

than the area of an individual ball. Two automobiles do not have to collide head on in order to suffer damage. Physicists collide atomic-sized particles with each other in big accelerators with ease and measure the cross section in units of **barns**, where 1 barn $= 10^{-28}$ (meter)^2. Collision cross sections are also measured in terms of $\pi(a_0)^2 \approx 0.88$ Å^2, where a_0 is the Bohr radius and the geometric size of an atom is comparable to 1 square angstrom (1 Å$^2 = 10^{-20}$ meter^2).

Example 3.1 Find the collision cross section of a collision if the radius of the target particle is r_T and the radius of the colliding particle is r_c.

Answer: As shown in the sketch, any particle that passes within $r_T + r_c$ of the target center will suffer a collision.

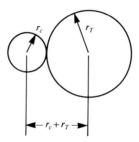

Hence the collision cross section is $\pi(r_c + r_T)^2$.

As the billiard ball passes through the array of other billiard balls, it may suffer more than one collision, as shown in Figure 3.2. The average distance

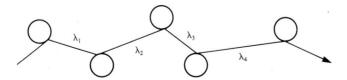

Figure 3.2 Path of a ball after colliding with other balls.

between collisions is called the **mean free path** λ_c, where

$$\lambda_c = \frac{1}{N} \sum_{i=1}^{N} \lambda_i \tag{1}$$

The average time between collisions is called the mean **collision time** τ_c.

$$\frac{\lambda_c}{\mathscr{U}} = \tau_c \tag{2}$$

where $\mathscr{U}$ is an average drift speed. The collision frequency v_c is defined as $1/(\text{collision time})$; $v_c = 1/\tau_c$.

Let us generalize this model to the case where there are N balls in a volume V; hence the density of the balls will be $n = N/V$. A ball enters the volume whose cross section is A with a velocity $\mathbf{v}$ as shown in Figure 3.3. In the slab within the volume whose thickness is Δx, the balls are arranged such that one ball does not hide behind another.

In the slab, there are ΔN balls, where

$$\Delta N = nA\,\Delta x$$

and the actual area covered by the balls that could have a collision is given by

$$\sigma_c\,\Delta N = \sigma_c nA\,\Delta x$$

The probability of a collision in Δx is given by

$$P\,\Delta x = \frac{\text{area covered by the balls}}{\text{area of the box}}$$

$$= \frac{\sigma_c nA\,\Delta x}{A} = \sigma_c n\,\Delta x \tag{3}$$

The probability is $P\,\Delta x$ since it depends directly on the size of the interval Δx.

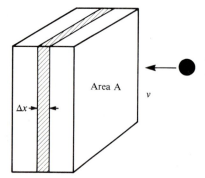

Area A

v

Δx

Figure 3.3 Ball colliding with a volume filled with N balls uniformly distributed in a volume V.

If N_0 balls enter the box and if they disappear after a collision in this region Δx, we can find the number that have been removed from the product of the density $N(x)$ of balls times the probability that there was a collision:

$$\Delta N(x) = -N(x)P \Delta x$$

or

$$\frac{\Delta N(x)}{N(x)} = -P \Delta x = -\sigma_c n \Delta x \tag{4}$$

The integral of (4) is

$$N = N_0 e^{-(x/\lambda_c)} \tag{5}$$

where $\lambda_c = 1/n\sigma_c$. The density of the balls that are incident on the volume decreases exponentially with a "space constant" given by the mean free path λ_c. The calculation presented here has just covered the behavior of hard spheres, which would correspond to billiard balls that disappear after a collision. There are several other mean free paths corresponding to different physical processes.

II. Transport Properties

Since we are considering nonequilibrium thermodynamical situations, it is possible to allow for the flow of particles under the influence of an external force. For example, it is possible to have a current under the application of an electric force. In this case, we can suggest the existence of **Ohm's law**, which is written as

$$\mathbf{j} = \sigma \mathbf{E} \tag{6}$$

where $\mathbf{j}$ is the current density (amperes/meters^2) and $\mathbf{E}$ is the electric field (volts/meter). The quantity σ is called the conductivity (mhos/meter or siemens/meter). Both are vector quantities since a direction is implied. The current is defined in terms of the nonequilibrium distribution function

$$\mathbf{j} = \frac{q}{m} \int \mathbf{p} f(\mathbf{r}, \mathbf{p}) \, d\mathbf{p} \tag{7}$$

which is not zero as it was in equilibrium situations.

The mobility μ (meter^2/volt-second) is defined from

$$\langle \mathbf{v} \rangle = \mu \mathbf{E} \tag{8}$$

where μ is positive for protons and negative for electrons. The average drift velocity is $\langle \mathbf{v} \rangle$. Therefore, the conductivity is related to the mobility via

$$\mathbf{j} = ne\langle \mathbf{v} \rangle = ne\mu \mathbf{E} = \sigma \mathbf{E}$$

TABLE 3.1

	σ (s/m)	μ (cm^2/V-sec)
Silver	6×10^7	10
Gallium arsenide	0.5	8500
Germanium	10	3900
Quartz	2×10^{-17}	—

from which we conclude that

$$\sigma = ne\mu \tag{9}$$

The conductivity and mobility have a wide range of values for different materials. Four are listed in Table 3.1, where the commonly employed units are used, and are applicable at room temperature. These materials represent three types of electrical materials: conductor, semiconductor, and insulator. More about this later.

Example 3.2   Given a resistor whose cross-sectional area is A and whose length is L, calculate the conductivity of the material in the resistor if only the normal electrical measurements of current and voltage are available.

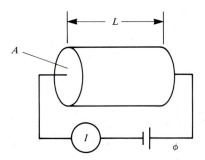

Answer: The conductivity σ is defined from $j = \sigma E$, where the vector directions are understood. The current density $j = I/A$ and the electric field is ϕ/L. Therefore, the conductivity $\sigma = j/E = (I/A)/(\phi/L)$.

Let us derive expressions for these quantities from a fairly simple model, that of an ionized gas where the density of the electrons n_e is equal to the density of ions n_i, and both densities are much less than the density of the background gas n. This is called a partially ionized plasma. Since the mass of the ion is at least 1836 times heavier than that of the electron and if the frequencies are sufficiently high, we can neglect the motion of the ions due to their inertia. Let us now inquire about the behavior of the electrons if the plasma is inserted in an electric field E.

As this is an external force on the particles, the electrons will obey Newton's law, which states that

$$m_e \frac{d\mathbf{v}}{dt} = -q\mathbf{E} \tag{10}$$

This equation is easily solved to yield that the electrons will be accelerated due to the electric field. These electrons will collide with a neutral particle, and due to the mass difference and conservation of momentum, their velocity just after collision will be ≈ 0. The electrons will again be accelerated to a velocity v_{max}, which can be computed from

$$v_{max} = \int_0^{\tau_c} \left(\frac{-qE}{m_e}\right) dt = -\frac{qE}{m_e}\tau_c$$

where τ_c is the average time between collisions. The average drift velocity of electrons $= v_{max}/2$ since the initial value was assumed to be zero. Therefore, the mobility is defined as

$$\mu_e = \frac{q\tau_c}{2m_e}$$

The conductivity σ is written as

$$\sigma = n_e q \mu_e = \frac{n_e q^2 \tau_c}{2m_e}$$

This will be derived with more rigor using a kinetic model later, and we will find that this simple model will be correct within a factor of 2.

III. Diffusion

The process of diffusion is one of the fundamental physical processes in nature. The diffusion equation describes the evolution of temperature in space and time from a hot source or to a cold sink. As such, the diffusion equation is sometimes called the heat equation. It can be used to describe the spreading of an aroma from a localized source or a signal passing along the nerve of a mammal.

      To derive the diffusion equation, we must first understand the equation of continuity. This can be derived in one dimension as follows. Consider, as shown in Figure 3.4, that a flux of particles per unit area $n\langle v \rangle$ is passing through the plane at x. Between this slab, which is located between the planes at x and $x + \Delta x$, is a distribution of ΔN particles. In this slab, $\Delta N = n\,\Delta x$ and n is the density of particles per unit area. The equation of continuity expresses the conservation of particles within the slab in that particles are neither created nor destroyed; they

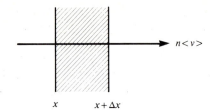

Figure 3.4 Flux $n\langle v \rangle$ passing through a slab of thickness Δx.

x $x + \Delta x$

can only enter or leave. Hence

$$\frac{\partial \, \Delta N}{\partial t} = \Delta x \frac{\partial n}{\partial t} = n\langle v \rangle \Big|_x - n\langle v \rangle \Big|_{x+\Delta x}$$

$$\Delta x \frac{\partial n}{\partial t} \approx n\langle v \rangle \Big|_x - \left[n\langle v \rangle \Big|_x + \frac{\partial n\langle v \rangle}{\partial x} \Big|_x \Delta x + \cdots \right]$$

Keeping only the lowest-order terms, this becomes

$$\frac{\partial n}{\partial t} + \frac{\partial (n\langle v \rangle)}{\partial x} = 0 \tag{11}$$

which is the equation of continuity. In three dimensions, the operator ∇ replaces the spatial derivative. In rectangular coordinates, this operator is

$$\nabla = \frac{\partial}{\partial x} + \frac{\partial}{\partial y} + \frac{\partial}{\partial z}$$

The flux $n\langle v \rangle$ can also be computed. The average drift velocity is taken to be $\langle v \rangle$. If there is a density gradient, the flux is found by evaluating the density at x and at $x + \Delta x$ such that

$$\Delta N \langle v \rangle = n \, \Delta x \langle v \rangle = D \left[n \Big|_x - n \Big|_{x+\Delta x} \right] \approx D \left[n \Big|_x - \left(n \Big|_x + \frac{\partial n}{\partial x} \Big|_x \Delta x + \cdots \right) \right]$$

which can be written as

$$n\langle v \rangle \approx -D \frac{\partial n}{\partial x} \tag{12}$$

Combine (11) and (12) and write

$$\frac{\partial n}{\partial x} - D \frac{\partial^2 n}{\partial x^2} = 0 \tag{13}$$

This is the diffusion equation and D is the diffusion coefficient. The units of D are $(\text{length})^2/\text{time}$. Many interesting problems arise if D is nonlinear, that is,

$D = D(n)$, or if D is spatially inhomogeneous, that is, $D = D(x)$. This will be encountered later in examining the transient behavior of a MOSFET.

Several formal procedures are available to solve the diffusion equation. The one that is the easiest is to follow in the footsteps of the giants who have trod the path ahead of us. Boltzmann suggested that the variables x and t should be combined such that

$$\Omega = \frac{x}{2\sqrt{Dt}} \tag{14}$$

Also, the dependent variable n should be replaced by

$$n(x, t) = t^r N(\Omega) \tag{15}$$

where r is a number that is to be later specified. The formal procedure for writing the combinations of (14) and (15) is based on seeking the invariants of certain Lie groups of algebra (see Appendix A). Substitute (14) and (15) in (13) and write

$$rt^{r-1}N(\Omega) + t^r \frac{dN(\Omega)}{d\Omega}\left(\frac{\partial\Omega}{\partial t}\right) - Dt^r \frac{d^2N(\Omega)}{d\Omega^2}\left(\frac{\partial\Omega}{\partial x}\right)^2 = 0$$

where the chain rule has been applied. Using (14), this equation becomes

$$rt^{r-1}N(\Omega) - \frac{1}{2}t^{r-1}\Omega\frac{dN(\Omega)}{d\Omega} - \frac{1}{4}t^{r-1}\frac{d^2N(\Omega)}{d\Omega^2} = 0$$

Note that the variables in time and space have disappeared since the factor t^{r-1} is common to all terms. Therefore,

$$rN(\Omega) - \frac{1}{2}\Omega\frac{dN(\Omega)}{d\Omega} - \frac{1}{4}\frac{d^2N(\Omega)}{d\Omega^2} = 0 \tag{16}$$

Let us specify the boundary and initial conditions that are applicable to the problem. The partial differential equation (13) corresponds to a third-order equation since there are three derivatives, two in space and one in time. Therefore, the physics of the problem must provide three conditions that must be satisfied in order to eliminate and/or specify the constants of integration. However, the second-order ordinary differential equation (16) requires only two boundary conditions or it will be overspecified. Two physical conditions, $n(x \to \infty, t)$ and $n(x, t = 0)$, *consolidate* to one condition at $\Omega \to \infty$. If both conditions were equal to zero, then we could write $N(\Omega \to \infty) = 0$. Another physical condition might be specified that

$$\int_{-\infty}^{\infty} n(x, t)\, dx$$

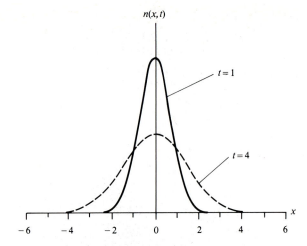

Figure 3.5 Solution of the diffusion equation.

is a constant or $n(x = 0, t)$ is equal to a constant. The first specifies that $r = -\frac{1}{2}$ and the second that $r = 0$. Choosing the first allows us to write (16) as

$$-\frac{1}{2}\frac{d[\Omega N(\Omega)]}{d\Omega} - \frac{1}{4}\frac{d^2 N(\Omega)}{d\Omega^2} = 0$$

This can be integrated to yield

$$-\Omega N(\Omega) - \frac{1}{2}\frac{dN(\Omega)}{d\Omega} = 0 \tag{17}$$

where the constant of integration is set equal to zero. The integral of (17) is

$$-\Omega^2 - \ln[N(\Omega)] = 0$$

where again the constant of integration is set equal to zero. Rewriting this with the inclusion of (14) and (15), we obtain

$$n(x, t) = \frac{n_0}{\sqrt{t}}e^{-x^2/4Dt} \tag{18}$$

This solution is shown in Figure 3.5 at various times after the introduction of the perturbing density. The density has diffused into the surrounding space.

Example 3.3   Show that (18) satisfies the requirement that

$$\int_{-\infty}^{\infty} n(x, t)\, dx$$

is a constant for all time.

Answer: The integral becomes

$$\int_{-\infty}^{\infty} \frac{n_0}{\sqrt{t}} e^{-x^2/4Dt} \, dx = 2\sqrt{D}\, n_0 \int_{-\infty}^{\infty} e^{-x^2/4Dt} \, d\left(\frac{x}{2\sqrt{Dt}}\right)$$

$$= 2\sqrt{D}\, n_0 \int_{-\infty}^{\infty} e^{-\beta^2} \, d\beta, \qquad \text{where } \beta = \sqrt{\frac{x^2}{4Dt}}$$

The integral can be performed and has the value $\sqrt{\pi}$. Therefore, the solution is $n_0(2\sqrt{D})\sqrt{\pi}$, which is a constant.

IV. Kinetic Equation

In the derivation of the Maxwell–Boltzmann distribution function in equilibrium thermodynamics, we made the assumption that there was spatial homogeneity and that there were no forces or fluxes in the system. In this section, an equation will be derived that will predict the evolution of a localized distribution into this special distribution. The derivation will follow the derivation of the equation of continuity, with the additional requirement that the localized distribution will also be localized in velocity v (or momentum p). This can be expressed as $f = f(x, v, t)$, and we will seek the evolution of f in *phase space* as shown in Figure 3.6.

The temporal evolution of the distribution function in the space $\Delta x \, \Delta v$ is given by

$$\frac{\partial f}{\partial t} \Delta x \, \Delta v = \Delta v \left\{ \langle v \rangle f \Big|_x - \langle v \rangle f \Big|_{x+\Delta x} \right\} + \Delta x \left\{ \left\langle \frac{F}{m} \right\rangle f \Big|_v - \left\langle \frac{F}{m} \right\rangle f \Big|_{v+\Delta v} \right\}$$

$$\approx \Delta x \, \Delta v \left\{ \frac{\partial(\langle v \rangle f)}{\partial x} + \frac{\partial(\langle F/m \rangle f)}{\partial v} \right\}$$

where a Taylor series expansion has been employed and $f\langle F/m \rangle$ corresponds to

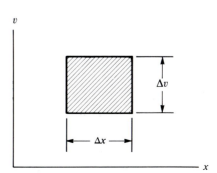

Figure 3.6   Phase space diagram.

the flux in momentum space. This is written as

$$\frac{\partial f}{\partial t} + \frac{\partial(\langle v \rangle f)}{\partial x} + \frac{\partial(\langle F/m \rangle f)}{\partial v} = \frac{df}{dt}\bigg|_{\text{collisions}} \tag{19}$$

where a collision term has been added to cause the distribution function to approach the Maxwell–Boltzmann distribution through irreversible processes. This is the kinetic equation that we seek. This equation is called the Boltzmann equation. There are several models for this collision term; the simplest is to set it equal to $v_c f$ and this is the one that will be studied here.

Another approach to derive the kinetic equation is to use the chain rule and find the total time derivative of $f(x, v, t)$, which is written as

$$\frac{df(x, v, t)}{dt} = \frac{\partial f}{\partial t} + \frac{\partial f}{\partial x}\frac{dx}{dt} + \frac{\partial f}{\partial v}\frac{dv}{dt} = \frac{df}{dt}\bigg|_{\text{collisions}}$$

The term dx/dt is the velocity v and the term dv/dt is the acceleration $a = F/m$. Therefore, the kinetic equation tells how the distribution function evolves in the two-dimensional phase space. In three dimensions, we deal with vectors and a six-dimensional phase space, and the one-dimensional spatial and velocity derivatives appearing in (19) become gradient operators in real and in velocity space, respectively.

It is possible to obtain solutions for the Boltzmann equation and one solution will be presented by an example.

Example 3.4   The kinetic equation (19) describes the evolution of a distribution function $f(x, v, t)$ in phase space. Describe the evolution of an initial distribution, which is shown in the figure, if there are no collisions or forces present.

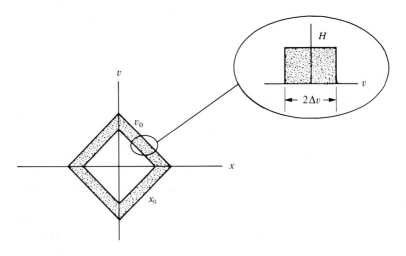

Answer: The density at this time, which will be taken as $t = 0$ at a particular location x where $-x_0 < x < x_0$, is given by

$$\int_{-\infty}^{\infty} f \, dv$$

Let the height of the distribution function at $t = 0$ be H. Therefore, we compute

$$N = \int_{-\infty}^{\infty} f \, dv = \frac{H}{2\,\Delta V} + \frac{H}{2\,\Delta V} = \frac{H}{\Delta V}$$

This density is constant for all x in the range $-x_0 < x < x_0$. The distribution function will evolve in space and time according to (19) as

$$\frac{\partial f}{\partial t} + v \frac{\partial f}{\partial x} \approx 0$$

since the collisions and the forces are assumed to be zero. The solution of this equation is given by $f(x, v, t > 0) = f(x - vt)$; that is, it propagates to increasing values of x with the velocity v that it had at $t = 0$. Note that some portions of the distribution function will not move since the velocity at $x = x_0$ is zero, while that at $x = 0$ has the maximum velocity $v = v_0$. Therefore, the distribution function will evolve as shown in the following figure:

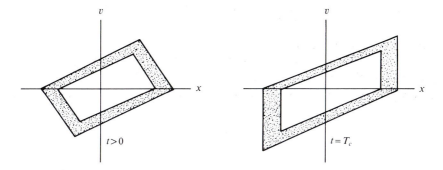

At a time called the critical time, the distribution function has steepened and the density will increase in value at that location and time. This critical time can be computed from $\partial f/\partial t = -\infty$.

Let us derive the mobility term with a little more rigor than given in Section II as an application of the use of the kinetic equation. A number of simplifications will have to be made. Even then, we will find that the obtaining of a solution will not be easy. The problem that will be treated is obtaining the evolution of an initial distribution to a Maxwell–Boltzmann distribution that is time independent and homogeneous. Furthermore, the collision term will be modeled as

$$\frac{df}{dt}\bigg|_{\text{collisions}} \approx \frac{f - f_{\text{MB}}}{\tau_c}$$

Therefore, we must solve

$$\frac{F}{m}\frac{\partial f}{\partial v} \approx -\frac{f - f_{\text{MB}}}{\tau_c}$$

or

$$f = f_{\text{MB}} - \tau_c \frac{F}{m}\frac{\partial f}{\partial v} \qquad (20)$$

Even this equation is difficult to solve. A serious and interested reader would attempt a solution using "variation of parameters." As a first approximation to obtain an intuition for the expected solution, let

$$\frac{\partial f}{\partial v} \approx \frac{\partial f_{\text{MB}}}{\partial v}$$

If this is done, then (20) becomes

$$f \approx f_{\text{MB}}\left[1 + \tau_c \frac{F}{m}\frac{mv}{k_B T}\right] \qquad (21)$$

where the Maxwell–Boltzmann distribution

$$f_{\text{MB}} = f_0 \exp\left(\frac{-mv^2}{2k_B T}\right)$$

is incorporated into the derivation. The average velocity is defined as

$$\langle v \rangle = \frac{\displaystyle\int_{-\infty}^{\infty} fv\, dv}{\displaystyle\int_{-\infty}^{\infty} f\, dv} \qquad (22)$$

where the normalization has been introduced. Substitute (21) in (22) and obtain two integrals in the numerator. The integral in the denominator leads to n_0, the density of particles. Also, recall that

$$\int_{-\infty}^{\infty} f_{\text{MB}} v\, dv = 0$$

Therefore, we finally obtain

$$\langle v \rangle \approx \frac{q}{k_B T} \langle v^2 \tau_c \rangle E \tag{23}$$

where the force $F = qE$. With $\langle v \rangle = \mu E$, we can identify the mobility μ as

$$\mu = \frac{q}{k_B T} \langle v^2 \tau_c \rangle$$

In a limited number of cases it is possible to assume that the average time between collisions τ_c is independent of the velocity. We can then write that

$$\mu \approx \frac{q \tau_c}{m} \tag{24}$$

In writing this, we have set the kinetic energy of the particles $mv^2/2$ equal to their thermal energy $k_B T/2$. This differs only by a factor of 2 from our previous result, which was obtained by elementary means. In three dimensions, the thermal energy is $3k_B T/2$.

V. Conclusion

Collisions are one of the mechanisms that allows energy to be dissipated such that a non-Maxwell–Boltzmann distribution can evolve into one. Waves will also dissipate energy from a particle distribution. The process of diffusion outlined here will be encountered later as it is an important physical technique for the actual manufacture of semiconductor devices.

The use of the kinetic equation gives the reader some comfort even at this stage of understanding. Several scores of Ph.D. theses and scholarly papers have been dedicated to the verification of the ramifications of kinetic theory. It appears to be on a firm foundation, and quantities correctly derived from it can be considered valid.

PROBLEMS

1. Show that the mean free path depends on the pressure P of the gas in which collisions take place. Use the definition for the equation of state $P = nk_B T$.

2. Using the formula derived in Problem 1, calculate the mean free path for collisions for argon at $P = 760$ torr and at $P = 1 \times 10^{-4}$ torr if the collision cross section is 20×10^{-20} meter^2 for an argon ion with 140 eV of energy. The answer for the lower-pressure case is approximately 1 meter, and this corresponds to a collision of

an argon ion with a neutral argon atom; the charge is exchanged during the collision such that the colliding ion becomes neutralized and the atom becomes ionized.

3. Show that the mean free path of an electron colliding with a gas consisting of two constituents that have collision cross sections σ_A and σ_B for electron interaction is given by

$$\frac{1}{\lambda} = \frac{1}{\lambda_A} + \frac{1}{\lambda_B}$$

where λ_A and λ_B are the mean free paths for electron collisions with the individual atom.

4. Find the number density of electrons in silver and germanium.

5. Sketch the solution for the diffusion equation given in (18) for three times: $T, 2T,$ and $3T$. Make sure that you have correctly labeled the axes of your graph.

6. Solve equation (20) using the "variation of parameters" method.

7. Derive equations (23) and (24).

8. In Example 3.4, find the time when the distribution function will have steepened and have an infinite slope. You will need to approximate the phase space distribution at $t = 0$ and watch how it evolves.

Chapter Four

Energy Levels

The detailed structure of an atom has been presented in the early chapters. If the reader is interested only in a deeper understanding of the structure of an atom, this book should now be closed. Our colleagues in elementary particle physics have written detailed accounts of their trials and tribulations using huge particle accelerators, which led to the discovery of a large number of new particles, a number that requires both the Roman and Greek alphabets to identify them. With new more powerful accelerators, we may expect that they'll start on the Hebrew or Babylonian alphabets next, since the first two are getting used up. However, if the reader is interested in making use of a more limited understanding of an individual atom, the following should be of interest.

Individual atoms are not of much practical use in building bridges, rockets, or transistors. Each solid object is comprised of atoms, but in a number that is huge (1 with tens of zeros following it). The energy level of each individual atom will align itself to that of its neighbor so that it can exist in harmony. In this chapter, we will study this process of lining up and be able to classify various materials as conductors, semiconductors, or insulators. Prior to this classification, the crystal structure of materials will be described.

I. Crystal Structure

If the world were a two-dimensional world, we could easily sketch the structure of materials by recalling the ideas from chemistry of bonding and drawing an elementary diagram of a material, say silicon, which has four electron in its outermost ($n = 3$) state. As shown in Table 1.2, a filled $n = 3$ state contains eight electrons. We might expect that the atoms would cooperate and share electrons with each other, and the two-dimensional atom would appear as the simplified structure shown in Figure 4.1, where the dashed lines indicate the sharing of electrons between the nearby atoms.

The world is, however, three dimensional, and thus the atoms arrange themselves in a three-dimensional structure called a crystal. The crystal forms because it is energetically favorable for it to do so. This is because there are forces of attraction and repulsion (hence attractive and repulsive potentials) between charges, and if the repulsive potential is inversely proportional to a higher power of r than the attractive potential, a local minimum of energy can exist. For

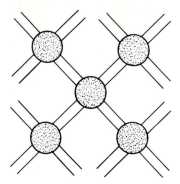

Figure 4.1 Simplified crystal structure consisting of atoms with four electrons in their outer shell.

example, the potential surrounding a single charge varies as $1/r$, while that about a dipole varies as $1/r^2$. Hence the total potential ϕ in the region of the sum of the charge and the dipole varies as

$$\phi = -\frac{A}{r} + \frac{B}{r^2}$$

where the first term corresponds to the attractive potential and the second to the repulsive potential. This is generalized to

$$\phi = -\frac{A}{r^m} + \frac{B}{r^n}, \qquad n > m \tag{1}$$

where n, m, A, and B are constants associated with the particular material. A sketch of this potential is shown in Figure 4.2 and the minimum at a radius $r = r_c$ is indicated.

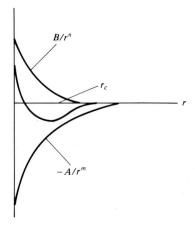

Figure 4.2 Potential variation surrounding a charged region consisting of two different types of charged configurations.

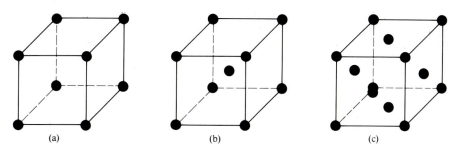

Figure 4.3   Three of the possible 14 Bravais lattices. (a) A simple cubic lattice with an atom located at each corner of the simple cell. (b) The body-centered cubic lattice has an additional atom at the center of the unit cell. (c) The face-centered cubic lattice has atoms at the eight corners of the cube and centered on the six faces.

The value of r_c represents the equilibrium spacing between atoms (or groups of atoms) for this particular crystal. It is an average value that can change due to changes in the ambient temperature or pressure. It could even undergo a radical change of phase and form a different crystal structure.

The attractive forces that balance the repulsive forces vary considerably both in origin and in magnitude for different materials. They are sometimes called short-range forces since the binding energy decays rapidly to zero. The weakest attractive force has a binding energy proportional to $1/r^6$. This force is called the *van der Waals* force. The Coulomb force or gravitational force is proportional to $1/r^2$, and it is called a long-range force.

The number of possible locations of these atoms or groups of atoms can be categorized into 14 possible different lattice structures, which are called the *Bravais lattices*. Three of the possible 14 lattice structures are shown in Figure 4.3.

Each Bravais lattice (each of the possible 14 is called a unit cell) can be characterized by three vectors that need not be orthogonal or equal in length. The quality that characterizes these lattices is that the arrangement and orientation of the array appear exactly the same when viewed from any point in the array.

With the arrangement of the atoms now known, it is possible to determine the percentage of the lattice that is actually filled with atoms. This will be demonstrated with the face-centered cubic lattice, which is shown in Figure 4.4, where the atoms are considered to be hard spheres.

The volume of each atom at each of the eight corners is distributed among the eight cubes that are joined at that corner. Hence each atom contributes one-eighth of its volume to a particular cube. The atom on each of the six sides contributes one-half of its volume to this cube. The number of atoms per cube is therefore

$$8\left(\frac{1}{8}\right) + 6\left(\frac{1}{2}\right) = 4$$

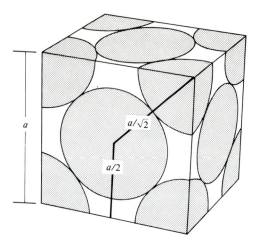

Figure 4.4   Packing of spheres in a face-centered cubic lattice.

Let the volume of the cube be $a \times a \times a$. The center of one face of the cube will be at $(a/2, a/2)$, so the distance from the center of a face and the corner is

$$\left[\left(\frac{a}{2}\right)^2 + \left(\frac{a}{2}\right)^2\right]^{1/2} = \frac{a\sqrt{2}}{2}$$

and the radius of the spheres is $a\sqrt{2}/4$. The volume of each sphere is

$$\frac{4\pi}{3}\left(\frac{a\sqrt{2}}{4}\right)^3 = \frac{\pi a^3 \sqrt{2}}{24}$$

The percentage of the cube that is actually filled with atoms is computed from

$$\frac{\text{Number of spheres} \times \text{Volume of each sphere}}{\text{Volume of cube}} = \frac{4 \times \left(\dfrac{\pi a^3 \sqrt{2}}{24}\right)}{a^3} = 74\%$$

This is a rather dense packing for the atoms with no spacing between each individual hard sphere atom. This assumption may not be very good in metals and covalent crystals.

Example 4.1   Calculate the percentage of a "simple cubic" atom that is filled with atoms. Assume that the volume of the lattice has a dimension $(2a) \times (2a) \times (2a)$, where a is the dimension of the atom.

Answer: Eight atoms are centered on the eight corners of the lattice, and each atom is shared among eight neighboring lattices. Therefore, the number of atoms in each cube is equal to

$$(8 \text{ corners}) \times \left(\frac{1}{8} \text{ atom per corner}\right) = 1 \text{ atom}$$

The volume of the atom is $4\pi a^3/3$, where a is the radius of the atom. The percentage of the lattice that is filled is

$$\frac{(4\pi a^3)/3}{(2a)^3} = \frac{\pi}{6} = 52\%$$

In general, the three types of crystalline structure can be divided into three classes: amorphous, polycrystalline, and single crystal. These three classes can be differentiated by the size of the ordered regions within the material. These range from a few molecular distances, to grains made up of highly ordered crystalline regions of irregular size and orientation, to a large, ordered, perfect single crystal. For applications, we find conductors possessing a fine-grain polycrystalline form. A semiconductor is usually a large single crystal with carefully controlled defects. Insulators can have any of the three forms. Materials found in nature usually have to be refined and processed before they can be employed in electrical devices. As X-ray diffraction techniques and electron-beam X-ray techniques improve, the detailed crystal structure of many of the materials are starting to be "seen" and modified to suit the application.

II. Energy Bands

The study of crystal structure will tell something about the mechanical properties of materials, such as malleability, hardness, and tensile strength. The electrical properties depend on the relative concentration and potential energy of the electrons. The two properties are related, although materials with similar mechanical properties can have vastly different electrical properties. The electrical properties are a sensitive function of the quantum interactions that determine the potential energy and momentum of electrons in the solid.

To illustrate a quantum interaction, let us first consider the interaction of joining two atoms, as shown in Figure 4.5. For just two atoms, there will be closely spaced discrete energy levels. We will let these levels be spread over a distinct energy spread in order to later introduce the idea of bands. In the figure, we have sketched the vertical coordinate in terms of the energy required to remove an electron from the atom; the maximum energy is required to remove the electrons that are closest to the positive nucleus. The discrete energy levels of each atom will adjust themselves so that they line up.

We can think of realigning in terms of a mechanical model as shown in Figure 4.6. In Figure 4.6(a), two masses are connected to the side walls with springs. The motion of one mass will be uncoupled from the motion of the other

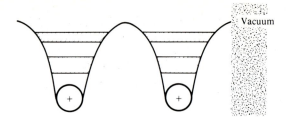

Figure 4.5   Two atoms in close juxtaposition have their energy levels aligned.

mass. However, as shown in Figure 4.6(b), if the masses are connected together with a spring, then the motion of one mass will influence the motion of the other mass. The two masses are coupled together.

For the case of the energy levels shown in Figure 4.5, the energy levels will adjust themselves due to a coupling between the adjacent atoms. The motion of an electron in one atom will influence the motion of the electrons in the adjacent atom. If there are N electrons, where N is a large number, there may be some overlap of the energy levels, which are called *bands*.

These bands are identified in the following manner. The topmost filled band at a temperature $T = 0$ is called the *valence band*. The band just above the valence band is called the *conduction band*. It is obvious that the Fermi energy should be between these two bands for an insulator, as shown in Figure 4.7. Having identified the bands, it is possible to classify materials. If the energy gap between the conduction band and the valence band is large ($\gg k_B T$), few electrons will have the energy to jump from the valence band to the conduction band. This material is called an *insulator*. If the Fermi energy is in the conduction band, then there will be electrons in the conduction band where they could easily move to the adjacent atom. This material is called a *conductor*. Finally, a *semiconductor* is defined as a material having the Fermi energy between the conduction band and the valence band but with a small energy gap. Typical energy gaps for an insulator, a semiconductor, and a conductor are 5 eV, 1 eV, and $\frac{1}{10}$ eV, respectively.

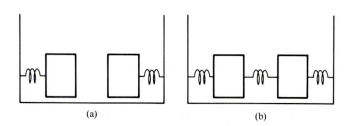

(a)   (b)

Figure 4.6   (a) Uncoupled and (b) Coupled spring–mass system.

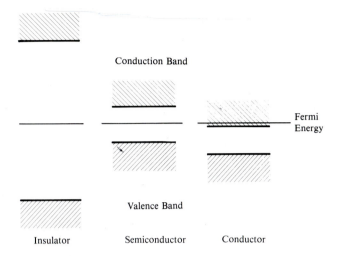

Figure 4.7   Energy level diagrams for the three types of materials.

III. Electrons in a Periodic Potential

The behavior of electrons in a periodic potential will determine their behavior in a crystal structure. Since the structure of a crystal is well defined and well ordered, we need only replace the crystal with a model consisting of a periodic potential, as shown in Figure 4.8. The electrons surrounding the nucleus have adjusted themselves, and we will just seek the behavior of one additional electron. A wave packet will be derived to describe this electron. If all electrons are statistically distributed through the single electron states and are assumed to behave like our chosen one, it is possible to predict characteristics of the material. One restriction must remain in the derivation; that is, the wave function $\psi(x, t) = X(x)T(t)$ that is derived must reduce to the wave function of a free particle,

$$X(x) = e^{ikx} \quad \text{or} \quad e^{-ikx} \tag{2}$$

as the depth of the periodic potential wells $\rightarrow 0$.

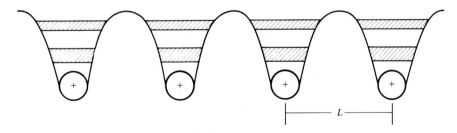

Figure 4.8   Periodic potential structure in a crystal.

Due to the periodicity requirement, it is advantageous to express the wave function in terms of a Bloch wave function defined by

$$X(x) = G(x)e^{ikx} \tag{3}$$

where $G(x) \to 1$ as the depth of the potential well $\to 0$. Periodicity of the system requires that

$$|X(x + nL)|^2 = |X(x)|^2$$

where L is the minimum translation in space from one lattice to the next. This distance L is called a basis vector. The wave number k can be specified to be

$$k = n\frac{2\pi}{L}$$

Remember that atoms are three-dimensional entities, although we will treat only a one-dimensional system for simplicity.

To find the wave function for the system, we have to solve the one-dimensional time-independent Schrödinger equation

$$\frac{d^2 X(x)}{dx^2} + \frac{2m}{\hbar^2}[E - U(x)]X(x) = 0 \tag{4}$$

Recall that E is the separation constant and $E = hv = \hbar\omega$. The problem that remains is to specify the periodic potential $U(x)$ such that (4) can be solved. If we choose it to be $U(x) = U_0 \cos ax$, then (4) can be solved in terms of Mathieu functions. This particular differential equation can also be used to describe the motion of a child's pumped swing. It has received considerable attention among mathematicians, and detailed studies of it have appeared. If the value for a is chosen to equal 2, which is twice the natural resonant frequency of (4), the equation will predict a parametric coupling of energy from a frequency 2Ω to a frequency Ω.

Although this potential is a reasonable approximation, the mathematics gets somewhat heavy. Therefore, a simpler approximation due to Kronig and Penney will be studied here. It will suffice to highlight the procedure and the underlying physics of a periodic potential structure.

IV. Kronig–Penney Model

The potential structure we outlined previously using a sinusoidal model for the periodic potential structure leads to a fairly complicated analysis. Kronig and Penney simplified the model by considering a periodic set of potential wells, as shown in Figure 4.9. In this case, the potential repeats itself at a distance of $x = L$.

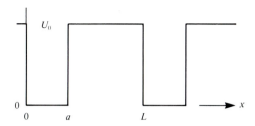

Figure 4.9   Periodic potential structure for the Kronig–Penney model.

The potential in one period can be expressed as

$$U(0 < x < a) = 0$$
$$U(a < x < L) = U_0 \tag{5}$$

The potential is periodic such that

$$U(x) = U(x + nL) \tag{6}$$

where n is an integer.

The procedure will be to substitute (5) into the one-dimensional time-independent Schrödinger wave equation (4) and solve it. Since the potential has two discrete values, we will have to separately solve the equations that are valid in the separate regions and connect the solutions by matching the boundary conditions. In particular, for the region $0 < x < a$, which will be called **I**, solve

$$\frac{d^2 X_{\mathrm{I}}(x)}{dx^2} + \frac{2m}{\hbar^2} E X_{\mathrm{I}}(x) = 0 \tag{7}$$

and in the region $a < x < L$, which will be called II, solve

$$\frac{d^2 X_{\mathrm{II}}(x)}{dx^2} + \frac{2m}{\hbar^2}(E - U_0) X_{\mathrm{II}}(x) = 0 \tag{8}$$

Recall the Bloch wave function (3). The second derivative of $X(x)$ can be expressed as

$$\frac{d^2 X(x)}{dx^2} = \left[\frac{d^2 G(x)}{dx^2} + 2ik\frac{dG(x)}{dx} - k^2 G(x) \right] e^{ikx}$$

where the subscript can be either I or II. Therefore, (7) becomes

$$\frac{d^2 G_{\mathrm{I}}(x)}{dx^2} + 2ik\frac{dG_{\mathrm{I}}(x)}{dx} + \left[\frac{2m}{\hbar^2} E - k^2 \right] G_{\mathrm{I}}(x) = 0 \tag{9}$$

Equation (8) can be written as

$$\frac{d^2 G_{II}(x)}{dx^2} + 2ik \frac{dG_{II}(x)}{dx} + \left[\frac{2m}{\hbar^2}(E - U_0) - k^2 \right] G_{II}(x) = 0 \tag{10}$$

Equations (9) and (10) can be easily solved. It is convenient to let

$$\beta_1 = \sqrt{\frac{2mE}{\hbar^2}}$$

and

$$\beta_{II} = \sqrt{\frac{2m(U_0 - E)}{\hbar^2}}$$

The solution of (9) is

$$G_1(x) = A e^{i(\beta_1 - k)x} + B e^{-i(\beta_1 - k)x} \tag{11}$$

and the solution of (10) is

$$G_{II}(x) = C e^{(\beta_{II} - ik)x} + D e^{-(\beta_{II} + ik)x} \tag{12}$$

The two solutions are connected smoothly at $x = 0$ and at $x = a$ (and at $x = L$ and at $x = -(L - a)$, etc., due to the periodicity). The "smoothness" of the connection implies that the functions and their derivatives are continuous at these points. Therefore,

$$G_1(0) = G_{II}(0)$$

$$\left. \frac{dG_1(x)}{dx} \right|_0 = \left. \frac{dG_{II}(x)}{dx} \right|_0$$

$$G_1(a) = G_{II}(a) \tag{13}$$

$$\left. \frac{dG_1(x)}{dx} \right|_a = \left. \frac{dG_{II}(x)}{dx} \right|_a$$

The simultaneous solution of these four conditions yields, after much algebra,

$$\left\{ \frac{\beta_{II}^2 - \beta_1^2}{2\beta_1\beta_{II}} \right\} \sinh[\beta_{II}(L - a)] \sin[\beta_1 a] + \cosh[\beta_{II}(L - a)] \cos[\beta_1 a] = \cos[kL] \tag{14}$$

This "dispersion relation" is a transcendental equation that is still difficult to solve as it stands.

However, it is possible to obtain an understanding of the solution in certain limits. This will give us a feeling for the meaning of the solution of this equation. The limit that we will examine is the limit of a deep wide potential well such that U_0 is large and $L - a$ is small, with the restriction that $U_0 a = $ constant. In this limit, (14) reduces to a simpler expression. To do this, we will examine each term in succession in (14). First, as $U_0 \to \infty$, $\beta_{II} \to \infty$ from the definition of β_{II}, so this limiting procedure becomes a limit of $\beta_{II} \to \infty$. Hence we find that the limiting expression is given by

$$M \left\{ \frac{\sin \beta_1 a}{\beta_1 a} \right\} + \cos \beta_1 a = \cos ka \qquad \textbf{(15)}$$

where

$$M = \frac{\beta_{II}^2 (L - a)a}{2}$$

There are only certain discrete values of β_1 that will satisfy (15). With β_1 defined as $\beta_1 = [2mE/\hbar^2]^{1/2}$, this means that there are only certain discrete values of energy that are allowed. In particular, if $M = 0$, then $\beta_1 a = ka$. If $M \to \infty$, then (15) can be satisfied only if $\beta_1 a = n\pi$, where n is an integer ($n = 1, 2, 3, \ldots$). From this, we compute the discrete energies that satisfy this equation as

$$E_n = \frac{(\pi \hbar n)^2}{2mL^2}$$

This is the same as a series of infinite potential wells and is valid for $E_n < U_0$.

For intermediate values of M, we must still solve (15) numerically. Prior to doing this, we note that the right side is limited to values between -1 and $+1$. The left side is shown in Figure 4.10 for a particular value of M. Note that for only a certain discrete set of $\beta_1 L$ will this equation have a value between -1 and $+1$.

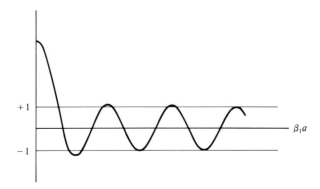

Figure 4.10   Solution of the left side of equation (15).

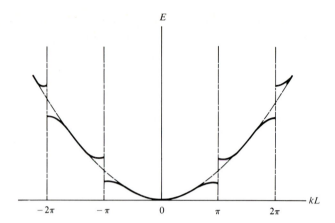

Figure 4.11   Brillouin zones.

For each value of energy E, there will be two values of k that satisfy (15) since the cosine function is an even function. A plot of E versus k for this equation is shown in Figure 4.11. The diagram is symmetric in k and there are discontinuities. These discontinuities in energy occur at the values $k = n\pi/L$. The ranges of k corresponding to the various energy bands are called *Brillouin zones*. They are sequentially numbered according to their satisfaction of the conditions

$$\frac{-\pi}{L} < k < \frac{\pi}{L}, \qquad \text{first Brillouin zone}$$

$$\frac{\pi}{L} < |k| < \frac{2\pi}{L}, \qquad \text{second Brillouin zone}$$

$$\frac{2\pi}{L} < |k| < \frac{3\pi}{L}, \qquad \text{third Brillouin zone}$$

and so on.

In each Brillouin zone, there are a large number of possible combinations of energy that can be used. Some of these are also indicated in Figure 4.11, with lighter lines where these additional quantities are confined to the first Brillouin zone. This reinforces the conjecture that there are only certain allowable energy states and that these states are discrete. This will become apparant in the next section, where these results will be interpreted.

V. Effective Mass

The interpretation of the electron behavior in a periodic potential well relies on our acceptance of the wave–particle duality that is at the foundation of quantum mechanics. Recall that in the derivation of the de Broglie wavelength the velocity of the particles was set equal to the group velocity of the wave in order to obtain the equivalent wavelength for the particles. In the obtaining of the solution of the Schrödinger wave equation for a periodic potential well as outlined previously, a fairly complicated dispersion relation resulted. The question that remains is if we can now reinterpret the particle behavior in terms of the wave behavior. Also, if such an interpretation is valid, what are its consequences?

To answer these questions, consider an electron moving in a crystal with a velocity v. This velocity will be assumed to be equal to the group velocity of the wave v_g, where

$$v_g = \frac{\partial \omega}{\partial k} = \frac{\partial (2\pi v)}{\partial k} = 2\pi \frac{\partial \left(\frac{E}{h}\right)}{\partial k} = \frac{1}{\hbar} \frac{\partial E}{\partial k} \tag{16}$$

The electron is allowed to pass through a localized region of electric field E, where it gains an increment of energy ΔE, where

$$\Delta E \approx -qE \, \Delta x$$

$$\approx -qEv_g \, \Delta t \tag{17}$$

This is shown in Figure 4.12.

Using the chain rule, the increment in energy for a wave can be expressed as

$$\Delta \mathscr{E} = \frac{\partial \mathscr{E}}{\partial k} \Delta k = v_g \hbar \, \Delta k \tag{18}$$

where (16) has been employed. Equating (17) and (18) with the assumption that the increments of gained energy for the particles are equal to that of the wave, we find

$$\frac{\partial (\hbar k)}{\partial t} = -qE \tag{19}$$

The acceleration of an electron as it passes through the localized electric field

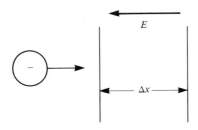

Figure 4.12   Electron passing through a localized electric field.

is given by

$$\frac{\partial v_g}{\partial t} = \frac{\partial\left(\frac{1}{\hbar}\frac{\partial\mathscr{E}}{\partial k}\right)}{dt} = \frac{1}{\hbar}\frac{\partial\left(\frac{\partial\mathscr{E}}{\partial t}\right)}{\partial k} = \frac{1}{\hbar^2}\frac{\partial\left(\frac{\partial\mathscr{E}}{\partial k}\frac{\partial(\hbar k)}{\partial t}\right)}{\partial k} = \frac{1}{\hbar^2}\frac{\partial^2\mathscr{E}}{\partial k^2}(-qE)$$

where (19) has been introduced. Therefore,

$$M^*\frac{\partial v_g}{\partial t} = -qE \tag{20}$$

where M^* is the effective mass of the wave and is defined as

$$M^* = \frac{\hbar^2}{\partial^2 E/\partial k^2} \tag{21}$$

This is in analogy to Newton's force equation of motion, where the force $(-qE)$ is equal to the mass times the acceleration. Note that if the derivative term is very large the effective mass will be small. This will result in a high mobility for the material.

The effective mass M^* can have a value that is greater than zero, equal to zero, or less than zero. The velocity of the particle is proportional to the derivative of the energy with respect to the wave number (or the crystal momentum). Both of these terms are local quantities as they both depend on the local value of k. For example, let $\mathscr{E} = -\mathscr{E}_0 \cos 2\pi k/k_0$ in the range $-\frac{1}{2} \le k/k_0 \le \frac{1}{2}$. At $k/k_0 = 0$, the mass is positive, while at $k/k_0 \approx \frac{1}{2}$, the mass is negative.

Values for the effective mass are computed in terms of the rest mass of an electron m_e. Typical average values at a temperature $T = 4°\text{K}$ for Ge, Si, and GaAs are found to have the following values:

	Ge	Si	GaAs
m_e^*	$0.082m_e$	$0.19m_e$	$0.067m_e$
m_h^*	$0.28m_e$	$0.49m_e$	$0.45m_e$

The concept of an effective mass has allowed us to incorporate the periodic potential of the crystal into the characteristics of the particle and think of it as a new particle that has some peculiar features. If it is positive, it will act as a normal particle in an external force field. If it is negative, its behavior will be the opposite. A particle with a positive mass will be accelerated in the direction of the external force field, while a particle with a negative mass will be decelerated.

VI. Conclusion

The study of the crystal properties of materials has led to several new ideas for the reader. The duality of waves and particles is emphasized, as we can take features from one domain and apply them to the other. The possible configurations for the crystal structure are limited to the 14 possible Bravais lattices due to energy and symmetry considerations. The atoms that occupy the lattice structure are compacted within the lattice but do not totally fill the volume. As the atoms are brought together, they couple with their neighbors and set up energy bands, and they lose their individual identities. The three types of materials, insulators, semiconductors, and conductors, are defined by the difference in energy levels between these bands. Finally, the effects of the individual atoms in a material can be neglected and replaced with a periodic potential structure.

PROBLEMS

1. Find the location of the potential minimum surrounding a single charge and a single dipole. Which must have an attractive or a repulsive potential? Assume that the magnitudes of the two charge configurations that are at the same location are equal.
2. Calculate the percentage of a body-centered cubic lattice that is filled.
3. Write the coupled differential equations that describe the spring–mass systems in Figure 4.6. You may assume that all masses are equal to M and all springs are defined by a spring constant k.
4. Calculate an energy difference at room temperature ($T = 300°K$).
5. Derive the equation for a child's swing with both parents "pumping" it, one on either side. Discuss what is happening in terms of "parametric coupling" of energy from 2Ω, which the parents are moving at, to Ω, which is the natural frequency of oscillation of

the child on the swing. Although each parent could move at a frequency Ω, you should assume that the parent will return to the "ready-to-push" position as the child passes the $\theta = 0$ location.

6. Find an electrical example that will yield the same effect as that described in Problem 5.

7. Derive Equation (14).

8. Derive Equation (15).

9. Obtain numerical results to correspond to the solution of (15). Choose values for the constants such that $M = 5$.

10. Discuss what would happen to the coupled mass sytem in Figure 4.6(b) if one of the masses were negative. What would happen to the energy?

11. Given that

$$\mathcal{E} = +\mathcal{E}_0\left(\frac{k}{k_0}\right)^2 \quad \text{or} \quad \mathcal{E} = -\mathcal{E}_0\left(\frac{k}{k_0}\right)^2$$

find the effective mass of a particle at $k = 0$.

12. Given that

$$\mathcal{E} = \mathcal{E}_0\cos\left(2\pi\frac{k}{k_0}\right)$$

find and sketch the value of the effective mass in the range $-2k_0 < k < 2k_0$.

Chapter Five

Semiconductors

The decades that followed the revolution that led to quantum mechanics can be thought of as a new enlightened era. The frontiers of space have been crossed, some diseases have been conquered, and powerful computers can be carried on the wrist. But as the mystery of the atom is being unveiled to the theorist and the experimentalist who work in large teams with large computers and accelerators, the realms of science fiction and reality start to overlap such that futurologists seem to be having schizophrenic fits in that they have either nothing to say or too much to say. The opening of the atom has had both its good and its bad aspects. Sociologists are having a heyday contemplating the isolation that people feel as they walk alone in a crowd with their ears plugged into an electronic mind manipulator. Perhaps the period since the development of the transistor should be thought of in terms of the words of Dickens, "It was the best of times, it was the worst of times...."

But as we have found after rubbing the magic lamp, it is hard to get the semiconductor genie back into the lamp. Since it is out, we should understand it and be able to control it. The physics that has preceded forms a necessary foundation. We are now ready to start building the first floor of our tower of knowledge. It is hoped that the reader's basis is such that another tower at Pisa will not result.

Semiconductors can be found in two varieties based on their purity. Those without any impurities are called intrinsic semiconductors and those that rely on an impurity are called extrinsic semiconductors. The details of each type will be explored in this chapter.

I. Intrinsic Semiconductors

Quantum mechanics has given us a prescription to calculate the number of vacant states that exist at various energy levels in a material. Quantum mechanics did not, however, tell us if the states were actually occupied. This was left to the Fermi–Dirac statistics, which predicted the actual occupation probability of a state. At room temperature, this function shows that most states in the conduction band will remain empty if the energy gap between the conduction band and the next lower band, which is the filled valence band, is greater than 1 electron volt.

Several materials do have an energy gap that is on the order of 1 electron volt. These materials formed the basis for the early semiconductor industry. They are given the name *intrinsic semiconductors* and comprise the following typical materials:

Material	Symbol	Energy gap (eV)
Gallium arsenide	GaAs	1.42
Germanium	Ge	0.66
Silicon	Si	1.12

The feature that further defines an intrinsic semiconductor is that, if an electron does appear in the conduction band, it must have come from the valence band, since that would be the only supply of electrons that could be excited into the conduction band. This is illustrated in Figure 5.1.

The probability that a state in the conduction band will actually be occupied is found from the Fermi function $f(E)$. The probability that a state in the valence band is unoccupied for an intrinsic semiconductor is also given by $f(E)$, since that electron has left the valence band and is now in the conduction band. The first Fermi function is valid in the range $E_c \leq E < \infty$, and the second applies to the range $E \leq E_v$.

The calculation of the number of electrons in the conduction band follows immediately. This, of course, is equal to the number of vacant states in the valence band by definition. These vacancies are called *holes*. As an electron in the valence band translates at the same energy level to the left to fill the hole created by the departed electron, it creates a new hole to its right. Hence there is an effective current of holes to the right. Recall that the valence band is positive, since one or more electrons have been promoted to the conduction band and this flow of holes is acting like the flow of positive charges. We can develop a mathematical description for this hole motion similar to the flow of electrons, and we will consider a hole to be a particle with a positive charge. By convention, the electrons are called *n*-type particles as they possess a negative charge and have an effective mass of $m_n^* = m_e^*$. The holes are called *p*-type particles since they act as positive charges and have an effective mass of $m_h^* = m_p^*$. The mass of either particle will be the effective mass that was described in Chapter 4.

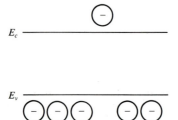

Figure 5.1   Intrinsic semiconductor in which one electron has been excited into the conduction band from the valence band.

The electrons in the conduction band will have an energy that is greater than the minimum energy of the conduction band E_c. This additional kinetic energy will be $p^2/2m_n^*$. Therefore, the total energy of these electrons will be

$$E = E_c + \frac{p^2}{2m_n^*} \tag{1}$$

Heisenberg's uncertainty principle states that the product of the uncertainty in knowing the particle's position and the uncertainty in knowing its momentum is greater than or equal to Planck's constant h. This is true in all three dimensions, so we can think of a volume in physical space and a volume in momentum space. Expressed in terms of symbols, we write

$$\Delta p_x \, \Delta x \, \Delta p_y \, \Delta y \, \Delta p_z \, \Delta z \geq h^3$$

Let us take the equality sign as the lowest limit and write

$$\Delta p_x \, \Delta p_y \, \Delta p_z = \frac{h^3}{V} \tag{2}$$

where $V = \Delta x \, \Delta y \, \Delta z$ is the spatial volume of the cube of uncertainty.

The number of states with momentums between p and $p + \Delta p$ that have a distribution $f(p)$ is computed from

$$g(p) \, \Delta p \approx \left(\frac{V}{h^3}\right) 2\Delta\left(\frac{4\pi}{3} p^3\right) \approx \left(\frac{V}{h^3}\right) 2(4\pi p^2 \, \Delta p) \tag{3}$$

In this equation, the factor of 2 accounts for the two directions of spin. The density of these states is given by

$$\frac{g(p)}{V} \Delta p$$

This can be converted into the more useful form of energy by employing (1) and writing

$$p = \sqrt{2m_n^*(E - E_c)}$$

Therefore

$$\frac{g(p)}{V} \Delta p = g_c(E) \, \Delta E = \left(\frac{1}{h^3}\right) 2(4\pi)(2m_n^*)^{3/2}[E - E_c]\left(\frac{1}{2}\right)\frac{1}{\sqrt{E - E_c}} \Delta E$$

or

$$g_c(E)\,\Delta E = 4\pi\left(\frac{2m_n^*}{h^2}\right)^{3/2}\sqrt{E - E_c}\,\Delta E \qquad (4)$$

The subscript c refers to the conduction band.

The number of electrons in the conduction band of an intrinsic semiconductor can be formally computed from

$$n_c = \int f(E)g_c\,dE \qquad (5)$$

where $f(E)$ is the Fermi–Dirac function. The limits of integration are from the bottom of the conduction band E_c to the top of the conduction band, which could be at $E \to \infty$. In practice, there is an upper limit to the conduction band, which will be taken as E_c'.

The number of holes in the valence band is computed in a similar fashion. We obtain

$$p_v = \int [1 - f(E)]g_v(E)\,dE \qquad (6)$$

where

$$g_v(E) = 4\pi\left(\frac{2m_p^*}{h^2}\right)^{3/2}\sqrt{E_v - E} \qquad (7)$$

In this integral, the limits are $E \to -\infty$ to the top of the valence band E_v. In practice, the minimum energy is at the bottom of the valence band E_v'.

Recall the definition of the Fermi–Dirac function. It is written as

$$f(E) = \frac{1}{1 + e^{(E - E_F)/k_B T}} \qquad (8)$$

If this is substituted into either (5) or (6), we find that the integral cannot be performed analytically except for the extreme limit of $T = 0$. However, we can approximate the Fermi–Dirac function at room temperatures to have a simpler form. At $T = 300$ K, the term $k_B T$ has a value of

$$(1.38 \times 10^{-23}\,\text{joule/K})\left(\frac{1}{1.6} \times 10^{-19}\,\text{eV/joule}\right)(300\text{ K}) = 0.026\text{ eV}$$

The unit of energy eV stands for electron volt and is equal to the energy that an electron would gain if it passed through a potential difference of 1 volt. For any material of interest in intrinsic semiconductors, the ratio of $(E_c - E_v)/k_B T \gg 1$.

Therefore, the exponential term in the Fermi–Dirac function is much larger than 1, and we approximate

$$f(E) \approx \exp\left(-\frac{E - E_F}{k_B T}\right) \tag{9}$$

Equation (5) therefore becomes

$$n_c = 4\pi\left(\frac{2m_n^*}{h^2}\right)^{3/2} \int \exp\left(-\frac{E - E_F}{k_B T}\right)\sqrt{E - E_c}\, dE$$

where the limits of integration are from E_c to the top of the conduction band. This can be rewritten as

$$n_c = 4\pi\left(\frac{2m_n^*}{h^2}\right)^{3/2} \exp\left(-\frac{E_c - E_F}{k_B T}\right) \int \exp\left(-\frac{E - E_c}{k_B T}\right)\sqrt{E - E_c}\, dE$$

The substitution of $\beta = (E - E_c)/k_B T$ converts the integral to

$$\int_0^{E_c} \sqrt{\beta}\, e^{-\beta}\, d\beta$$

with the limits of the integral being 0 and the top of the conduction band. If this limit is taken to be ∞, the value of the integral is given by $\sqrt{\pi/2}$. Therefore, the number of electrons in the conduction band is given by

$$n_c = 2\left(\frac{2\pi m_n^* k_B T}{h^2}\right)^{3/2} \exp\left(-\frac{E_c - E_F}{k_B T}\right) \tag{10}$$

It is common to define the effective density of states as

$$N_c = 2\left(\frac{2\pi m_n^* k_B T}{h^2}\right)^{3/2} \tag{11}$$

and therefore

$$n_c = N_c \exp\left(-\frac{E_c - E_F}{k_B T}\right) \tag{12}$$

In a similar manner, we find for the holes in the valence band that

$$p_v = N_v \exp\left(-\frac{E_F - E_v}{k_B T}\right) \tag{13}$$

where

$$N_v = 2\left(\frac{2\pi m_p^* k_B T}{h^2}\right)^{3/2} \tag{14}$$

and the effective mass of the holes is written as m_p^*.

This calculation is valid for an intrinsic semiconductor where, by definition, the number of electrons in the conduction band is equal to the number of holes in the valence band. Since they are equal, it is possible to find the value of the Fermi energy by setting $n_c = p_v$ from (12) and (13).

$$N_c \exp\left(-\frac{E_c - E_F}{k_B T}\right) = N_v \exp\left(-\frac{E_F - E_v}{k_B T}\right)$$

From this, we compute that

$$E_F = \frac{E_c + E_v}{2} + \frac{k_B T}{2} \ln\left(\frac{N_v}{N_c}\right) \tag{15}$$

Since N_c and N_v differ only by their effective masses, this can be written as

$$E_F = \frac{E_c + E_v}{2} + \frac{3k_B T}{4} \ln\left(\frac{m_p^*}{m_n^*}\right) \tag{16}$$

If the effective masses are equal or if the temperature is at $T = 0$ K, then the Fermi energy will lie in the middle of the gap between the conduction band and the valence band. In silicon at room temperature, m_n^* is slightly greater than m_p^*, so the Fermi energy lies a little below the middle.

Example 5.1 Calculate the deviation in the Fermi energy about its equilibrium position if the effective mass of the holes changes from an equilibrium value.

Answer: From equation (16), we write

$$E_F = \frac{E_c + E_v}{2} + \frac{3k_B T}{4} \ln\left(\frac{m_p^*}{m_n^*}\right)$$

Therefore, the deviation of the Fermi energy ΔE_F is given by

$$\Delta E_F = \frac{3k_B T}{4} \left(\frac{1}{(m_p^*/m_n^*)}\right)\left\{\Delta\left(\frac{m_p^*}{m_n^*}\right)\right\}$$

$$= \frac{3k_B T}{4} \frac{\Delta m_p^*}{m_p^*}$$

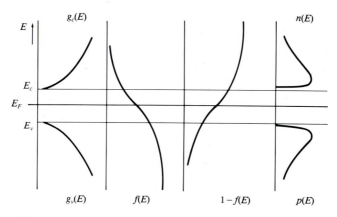

Figure 5.2   Distribution function of states $g_c(E)$ and $g_v(E)$ in the conduction and valence bands, respectively. The Fermi functions $f(E)$ and $(1 - f(E))$ allow us to find the electron and hole densities.

In Figure 5.2, the distribution of states, the relevant Fermi–Dirac functions, and densities of electrons in the conduction band and holes in the valence band are plotted as a function of energy. Note that the particle densities are concentrated close to the valence and conduction energies, respectively.

Example 5.2   Show that the probability that a state ΔE above the Fermi energy is occupied is equal to the probability that a state ΔE beneath the Fermi energy is unoccupied.

Answer: The probability that a state is occupied is given by the Fermi function $f(E)$, which is defined by

$$f(E) = \frac{1}{1 + e^{(E - E_F)/k_B T}}$$

Therefore, with $E = E_F + \Delta E$, we have

$$f(E_F + \Delta E) = \frac{1}{1 + e^{(E_F + \Delta E - E_F)/k_B T}} = \frac{1}{1 + e^{\Delta E/k_B T}}$$

The probability that a state is not occupied is given by $1 - f(E)$. Therefore,

$$1 - f(E_F - \Delta E) = 1 - \frac{1}{1 + e^{(E_F - \Delta E - E_F)/k_B T}} = 1 - \frac{1}{1 + e^{(-\Delta E)/k_B T}}$$

$$= \frac{1}{1 + e^{(\Delta E)/k_B T}}$$

The carrier concentrations in the bands are equal by definition. Therefore, we can write

$$n_c p_v = n_c^2 = N_c N_v \exp\left(-\frac{E_c - E_v}{k_B T}\right) \tag{17}$$

which allows us to compute the temperature dependence of n_c on temperature, since both N_c and N_v depend on temperature from (11) and (14). Although each quantity depends on temperature as $T^{3/2}$, the exponential term dominates. The carrier density also depends on the energy gap $E_c - E_v$, which is an intrinsic characteristic of the material. If this gap is larger, the intrinsic carrier concentration will decrease at a constant temperature.

The reader can imagine a very practical limitation for the manufacture and employment of intrinsic semiconductors. That, of course, is the stringent requirement concerning the allowed impurity level in the manufacturing process. A typical impurity content of just a few parts in 1 billion causes a disaster in the intrinsic semiconductor business. Small wonder that the employees involved in the manufacture dress in surgical caps and gowns. An "infection" could spell a depression.

Once an intrinsic semiconductor is actually manufactured, a current can be made to flow using the following arguments. The current flow in a filled band is equal to zero.

$$\mathbf{j} = \sum_{i=1}^{N} (-q)\mathbf{v_i} = 0 \tag{18}$$

where $\mathbf{j}$ is the current density, $\mathbf{v_i}$ is the velocity of the ith electron, and N is the concentration of electrons in the full band. If $n = p$ electrons have been promoted from the valence band to the conduction band, which is certainly not filled, these electrons can move, setting up a current defined by

$$\mathbf{j} = \sum_{i=1}^{n} (-q)\mathbf{v_i} = 0 \tag{19}$$

where n electrons are in the conduction band.

There can now be a current in the valence band since n electrons have left. This is given by

$$\mathbf{j} = \sum_{i=n+1}^{N} (-q)\mathbf{v_i} = 0 \tag{20}$$

From (18), we write

$$\sum_{i=1}^{n} (-q)\mathbf{v_i} + \sum_{i=n+1}^{N} (-q)\mathbf{v_i} = 0$$

Hence (20) can be written as

$$\mathbf{j} = \sum_{i=1}^{n} (+q)\mathbf{v_i} \tag{21}$$

which is equal to the current carried by the holes in the valence band. We have a choice in interpreting the current. A choice will be made to follow the convention that the carrier type is the less populated one in a given band. The motion of the particles will be caused by external electric and magnetic fields as in a conductor.

II. Extrinsic Semiconductors

In the manufacture of the intrinsic semiconductors just described, it was determined that a very small amount of impurities would have disastrous consequences for the ultimate success of the intrinsic semiconductor business. Is this true for the entire semiconductor business or is there another type of semiconductor that can be somehow based on the fact of life that impurities do exist? The answer to this question has led to a new industry, which is called the extrinsic semiconductor business. The surgical caps and gowns are still worn by the manufacturers, but the impurity content may be large. The impurity content is, however, of a known and specified variety within a known and specified host environment. The limits imposed by the cleanliness are still the same!

A typical and probably the most common host environment would be pure silicon. It contains four electrons in its outer shells. It has the propensity to share these four electrons with four of its neighbors and to share one of their electrons to form a stable crystal structure, as shown in Figure 5.3. Although stable, such a material is useful only as an insulator as it stands.

If now one of the silicon atoms is replaced with an impurity such as a gallium atom (an acceptor), which contains only three electrons in its outer shell, we find that there are not enough electrons to share with the neighboring silicon atoms. Similarly, if an impurity atom such as an arsenic atom (a donor) replaces a silicon

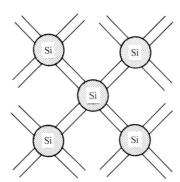

Figure 5.3   Silicon crystal structure.

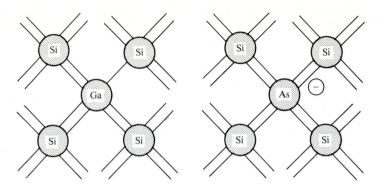

Figure 5.4 Extrinsic semiconductor with either gallium or arsenic dopants.

atom, we find that there is an extra electron available in the semiconductor. This
is shown schematically for these two cases in Figure 5.4.

The actual process of introducing the impurity into the host environment is
called *doping* the semiconductor, and the impurity is called the *dopant*. The
materials that can yield an extra electron are called *donors* and form an *n*-type
semiconductor, since there will be more electrons in the material than holes. For
this material, the electrons form the dominant carrier of charge so they would be
called the *majority* carrier. The holes would be the *minority* carrier.

For notational convenience, the superscript $^+$ indicates a heavily doped
semiconductor. Therefore, we will encounter the n^+ and p^+ notation. The $^-$
superscript is sometimes used to indicate light doping. The materials that can
absorb an extra electron due to a deficiency are called *acceptors* and form a *p*-type
semiconductor, since there are fewer electrons than holes. Recall that the motion
of an electron in one direction to fill a vacancy implies a hole current in the
opposite direction. The holes in this material would be the majority carrier and
the electrons would be the minority carrier.

The choice of which impurity to introduce into the host crystal is based on
the following considerations. For a final *n*-type semiconductor, the impurity must
have an energy band that contains electrons that are close to the conduction band
of the host crystal. The Fermi energy would be between this impurity level and
the conduction band and hence would be far from the valence band. Similarly, for
a *p*-type semiconductor, the impurity must possess an unfilled band that is close
to the valence band of the host crystal. The Fermi energy would then be closer to

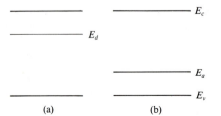

(a)   (b)

Figure 5.5 Two extrinsic
semiconductors: (a) *n* type; (b) *p* type.

the valence band. These energy levels will be denoted as E_d and E_a, respectively, with the subscripts chosen to reflect whether the band donates or accepts electrons. This is shown in Figure 5.5.

Let us now put some mathematics into the description of an extrinsic semiconductor. We will choose to examine an *n*-type semiconductor whose energy level diagram is shown in Figure 5.6.

The following definitions are employed for the densities of particles (usually number per cubic centimeter).

$$N_d = \text{donor atoms}$$

$$N_{do} = \text{nonionized donor atoms}$$

$$p = \text{holes in the valence band}$$

$$n = \text{electrons in the conduction band}$$

Charge neutrality requires that

$$n = p + (N_d - N_{do})$$
$$\approx (N_d - N_{do}) \tag{22}$$

since we are talking about an *n*-type semiconductor. For this case, we have placed the Fermi level between the donor impurity level and the conduction band. Hence we would expect that very few electrons from the valence band would actually be able to make it to the conduction band. The density of nonionized donor atoms is related to the number of donor atoms by the Fermi function.

$$N_{do} = f(E_d)N_d \tag{23}$$

This states that the number of filled states equals the probability that a state is occupied times the number of states. Therefore,

$$n = N_d(1 - f(E_d)) \tag{24}$$

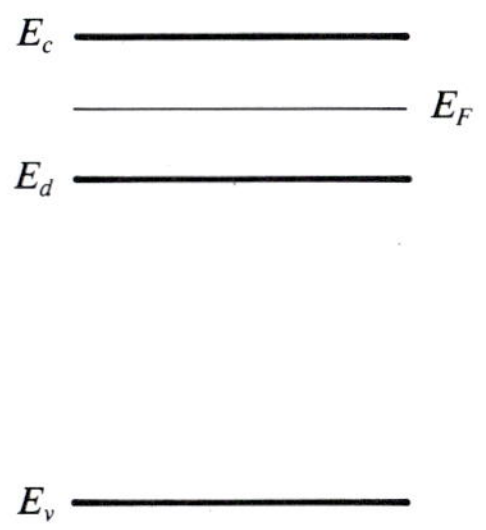

E_c ———————

——————— E_F

E_d ———————

Figure 5.6 Energy levels of an *n*-type
semiconductor. E_v ———————

Substitute the Fermi function in (24) and obtain

$$n = N_d \left\{ 1 - \frac{1}{1 + e^{(E_d - E_F)/k_B T}} \right\}$$

This can be rewritten as

$$n = \frac{N_d}{e^{(E_F - E_d)/k_B T} + 1}$$

which at low temperatures becomes

$$n \approx N_d \exp\left(-\frac{E_F - E_d}{k_B T} \right) \tag{25}$$

Let us set this equal to the number of electrons that are in the conduction band that is given by

$$n_c = 2 \left(\frac{2\pi m_n^* k_B T}{h^2} \right)^{3/2} \exp\left(-\frac{E_c - E_F}{k_B T} \right) \tag{26}$$

If this is done, the Fermi energy is given by

$$E_F = \frac{E_c + E_d}{2} + \frac{k_B T}{2} \ln\left\{ \frac{N_d}{2[(2\pi m_n^* k_B T)/h^2]^{3/2}} \right\} \tag{27}$$

If the temperature increases to very large values, we would expect that all the donor atoms would be ionized and $n \approx N_d$. In this case, the extrinsic semiconductor starts to behave as an intrinsic semiconductor. In fact, it is possible to show that the extrinsic densities can be written in terms of the intrinsic densities. We will do this with the following notational change. The subscripts e and i will be added to distinguish the extrinsic from the intrinsic variables. In particular, we write

$$n_e = n_i \exp\left[\frac{E_{Fe} - E_{Fi}}{k_B T} \right] \tag{28}$$

$$p_e = p_i \exp\left[\frac{E_{Fe} - E_{Fi}}{k_B T} \right] \tag{29}$$

To prove this, we make use of the known property of intrinsic semiconductors that

$$n_i p_i = N_c N_v \exp\left[-\frac{E_c - E_v}{k_B T} \right]$$

where

$$N_c = 2\left(\frac{2\pi m_n^* k_B T}{h^2}\right)^{3/2}$$

and

$$N_v = 2\left(\frac{2\pi m_p^* k_B T}{h^2}\right)^{3/2}$$

Also, the Fermi energy for the intrinsic semiconductor is given by

$$E_{Fi} = \frac{E_c + E_v}{2} + \frac{3k_B T}{4} \ln\left(\frac{m_p^*}{m_n^*}\right)$$

Substitute these terms into (28) and obtain

$$n_e = n_i \exp\left(\frac{E_{Fe} - E_{Fi}}{k_B T}\right)$$

$$= \left[N_c N_v \exp\left(-\frac{E_c - E_v}{k_B T}\right)\right]^{1/2} \exp\left(\frac{E_{Fe}}{k_B T}\right) \exp\left[-\frac{\frac{E_c + E_v}{2} + \frac{3k_B T}{4} \ln\left(\frac{m_p^*}{m_n^*}\right)}{k_B T}\right]$$

$$= [N_c N_v]^{1/2} \exp\left[\frac{E_{Fe} - E_c}{k_B T}\right]\left(\frac{m_p^*}{m_n^*}\right)^{-3/4}$$

$$= N_c \exp\left[\frac{E_{Fe} - E_c}{k_B T}\right]$$

where the relation that $N_v/m_p^{*3/2} = N_c/m_n^{*3/2}$ has been employed. This is the desired result, which is given by (12). A similar treatment of (29) would lead to (13).

In the process of manufacturing an extrinsic semiconductor, there is a possibility that a small amount of the wrong type of impurity can get mixed in with the desired dopant. All is not lost, since the donor and acceptor levels obey the occupation statistics that determine the conduction and the valence bands. Therefore, if N_d is greater than N_a, the Fermi energy will be closer to the conduction than the valence band. All acceptor states will be occupied with electrons from the donor states, and the material will act as if it were doped with $N_d - N_a$ donors. The opposite condition holds true if N_a is greater than N_d.

One of nature's laws comes into the discussion at this point, the second law of thermodynamics. Both the electrons and the holes will relax to the lowest energy level that they can find. This is equivalent to stating that the entropy of the system will increase, and we draw the analogy of a ball rolling up and down a hill until it comes to rest at the bottom of the valley. The ball has reached its minimum energy, but in the process it has heated up the hill and the air. The entropy of the system has increased. Therefore, donor electrons will drop and fill the acceptor states. The excess donors act as a net dopant and readily thermalize.

III. Optical Generation of Hole–Electron Pairs

The mechanism for the generation of hole–electron pairs described for the intrinsic and the extrinsic semiconductors relied on the temperature of the semiconductor's being nonzero. The Fermi–Dirac statistics then predicted that the states would be occupied. This would be a thermal generation mechanism for the creation of the charged particles. We would be justified in wondering at this point if there is another possible mechanism that could be employed for the creation mechanism. The answer to this thought is that there is another mechanism and it is based on the photoelectric effect described by Einstein.

Recall in our discussion of the underlying physics that led to the development of quantum mechanics that an electron could gain or lose energy if it changed orbits, but its energy would remain the same if it stayed in the same orbit. This was expressed mathematically with Einstein's equation for the photoelectric effect as

$$E_2 - E_1 = h\nu \tag{30}$$

An interpretation can now be given in terms of semiconductors, where we will let $E_2 = E_c$ and $E_1 = E_v$ for an intrinsic semiconductor. For an extrinsic semiconductor, we will replace these terms with the appropriate quantities of E_d or E_a.

One additional constraint must be included in the analysis. The absorption of a photon of light and the transition from an initial state to a final state must conserve both energy and momentum. We find that the momentum of the photon is negligible in comparison with the effective momentum of an electron in a crystal. There are cases where the momentum $\hbar \mathbf{k}$ changes during the transition, and these are called an *indirect transition*. If the momentum does not change during the transition, it is called a *direct transition*. Both types of transition are illustrated in Figure 5.7. The transition for GaAs is direct, while the transitions for

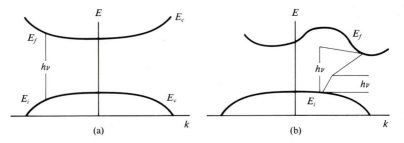

Figure 5.7 (a) Direct and (b) indirect transitions between the conduction and valence band.

Ge and Si are indirect. For the case of the indirect transition, the photon could be either created or destroyed, and hence there are two possible paths for the transition.

Let us assume that the energy of the photon is E_p, which is either created or destroyed in this process. Therefore, conservation of energy implies that

$$E_f - E_i(+ \text{ or } -)E_p = h\nu_{(+ \text{ or } -)} \tag{31}$$

where E_i and E_f correspond to the initial and final energy states. The details of calculating the probabilities of a specific transition are left to the area of quantum electrodynamics.

This type of electron–hole creation has generated strong interest in the solar cell form of electric power generation. It has found wide use in space applications and in solar cell watches and calculators. Several compounds exhibit the effect. For example, one could use gallium phosphide (GaP, $E_{gap} = 2.26$ eV), cadmium sulfide (CdS, $E_{gap} = 2.42$ eV) or silicon or germanium compounds doped with gold or mercury. From the experiments cited in Chapter 1, we note that these energy gaps are either in the visible range of wavelengths or close to it.

IV. Recombination and Lifetimes

In thermal equilibrium, an equilibrium concentration of electrons will reside in the conduction band and holes in the valence band of the semiconductor. If now a perturbation occurs in the system, such as a photon being momentarily incident upon the semiconductor, or a sudden change of temperature occurs, we might expect that a pertubation would occur in the equilibrium electron density in the conduction band. There are two scenarios that can result and affect this perturbation of the density. The first is that the electron could migrate away in the conduction band under the influence of an external electric field. This will be discussed in the next section. The second effect is that the electron could return to the valence band and *recombine* with a hole. It is this effect that will be examined here.

Let us assume that holes are generated at a rate g per unit time and recombine at a rate r per unit time. The one-dimensional equation of continuity is written as

$$\frac{\partial p}{\partial t} + \frac{\partial (p\langle v \rangle)}{\partial x} = g - r \tag{32}$$

where $\langle v \rangle$ is the average velocity of the holes. In thermal equilibrium, $g = r$, $\langle v \rangle = 0$, and $p = p_0$, which is a constant. As a first approximation, let us assume that the recombination rate r is proportional to the density of holes that exist at a particular time. For dimensional reasons, this proportionality constant must have the units of 1/time. Therefore, we write

$$r = \frac{p}{\tau_h}$$

Furthermore, we will assume that the source of hole generation is turned off and that $\langle v \rangle \approx 0$. Hence the excess charge perturbation $(p - p_0)$ will decrease according to (32), where p_0 is the new equilibrium value. The solution of (32) is

$$(p - p_0) = (p - p_0)\bigg|_{t=0} e^{-t/\tau_h} \qquad (33)$$

Note that the density decreases in time with a time constant τ_h. This time is called a *lifetime* or a *recombination time*.

This lifetime is extremely important in that it can determine a low-frequency cutoff for a semiconductor device. For frequencies $f < 1/\tau_h$, recombination has occurred and the device is ineffective for operation. As will be noted in the next section, recombination will be important when describing the transverse migration of charges. A similar lifetime τ_e can also be found for electrons.

V. Transport Coefficients

In the discussion of nonequilibrium thermodynamics, we discussed transport coefficients in general. The reader can probably speculate that similar coefficients should also be uncovered for semiconductors, and this indeed is the case. However, in this case there are two species of particles that can carry the charge. This will have an effect on the overall electrical characteristics of the semiconductor, as outlined in the following.

We can first estimate the random thermal velocity of the electrons in an n-type semiconductor by equating the electron's kinetic energy to the thermal energy, which is given by

$$\frac{m_e^* v_{th}^2}{2} = \frac{3k_B T_e}{2} \qquad (34)$$

where the effective mass has been employed. Also, there are three degrees of freedom for the thermal motion. An estimate of this velocity at room temperature for silicon and gallium arsenide yields a maximum thermal velocity of $v_{th} \approx 10^7$ cm/sec. These electrons are moving rapidly but in all directions. On the average, there is no net current since there is no preferred direction of electron motion.

However, if an electric field E is applied to the semiconductor, the electrons will have a directed velocity v_n, which can be written as

$$v_n = -\mu_n E \qquad (35)$$

where μ_n is the mobility of the electrons (cm^2/volt-second) and the vector direction is understood. The derivation for the mobility is identical to that described previously and will not be repeated here. A major difference for these electrons is that there are many different "objects" with which they can collide. For example, they can collide with an impurity in the lattice or with the lattice itself.

The holes will undergo a similar directed motion under the influence of an applied electric field, and we can write

$$v_p = \mu_p E \tag{36}$$

The total current density j is given by

$$j = j_n + j_p = -nqv_n + pqv_p = (nq\mu_n + pq\mu_p)E \tag{37}$$

where q is the charge of the individual particle. The densities for each type of particle are n and p. The *conductivity* σ of the semiconductor is given by

$$\sigma = (nq\mu_n + pq\mu_p) \tag{38}$$

Example 5.3   Find the conductivity of intrinsic Ge at room temperature.

Answer: For an intrinsic semiconductor, $n_i = p_i$ and

$$n_i = (N_c N_v)^{1/2} e^{-E_g/2k_B T}$$

The effective masses for electrons and holes in Ge are found to be $m_n^* = 0.55 m_e$ and $0.37 m_e$, respectively, where m_e is the electron mass. The energy gap E_g for Ge is 0.67 eV. Using (11) and (14), we write

$$N_c = 2\left\{\frac{2\pi(0.55 \times 9.1 \times 10^{-31} \text{ kg})(1.38 \times 10^{-23} \text{ J/K})(300 \text{ K})}{(6.6 \times 10^{-34} \text{ J-sec})^2}\right\}^{3/2}$$

$$= 1 \times 10^{19} \text{ cm}^{-3}$$

$$N_v = 2\left\{\frac{2\pi(0.37 \times 9.1 \times 10^{-31} \text{ kg})(1.38 \times 10^{-23} \text{ J/K})(300 \text{ K})}{(6.6 \times 10^{-34} \text{ J-sec})^2}\right\}^{3/2}$$

$$= 0.6 \times 10^{19} \text{ cm}^{-3}$$

Therefore,

$$n_i = [(1 \times 10^{19})(0.6 \times 10^{19})]^{1/2} \exp\left(-\frac{0.67 \text{ eV}}{2 \times 0.0259 \text{ eV}}\right) = 1.9 \times 10^{13} \text{ cm}^{-3}$$

The conductivity σ is computed from (38):

$$\sigma = (1.9 \times 10^{13} \text{ cm}^{-3}) \times (1.6 \times 10^{-19} \text{ coulomb})$$

$$\times ([3900 + 1900] \text{ cm}^2/\text{V-sec}) \approx 2 \times 10^{-2} (\text{ohm-cm})^{-1}$$

A common technique to measure the conductivity of a semiconductor is to use four equally spaced probes, as shown in Figure 5.8. A constant current source is connected to the outer two probes and the voltage is measured between the inner two. The current between the outer two probes will spread over the entire volume of the semiconductor. As the problem of finding the distribution of the electric field, and hence current distribution, between the two outer conductors is a problem that can be treated, we can measure the conductivity from

$$\frac{1}{\sigma} = \frac{\Delta\phi}{I} W \times (\text{form factor}) \tag{39}$$

where the form factor takes into account the spreading of the current.

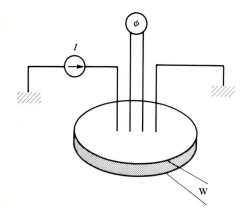

Figure 5.8   Technique to measure the conductivity of a semiconductor.

Example 5.4   Calculate the form factor for an experimental setup consisting of two large parallel plates and two small parallel plate probes as shown in the following figure:

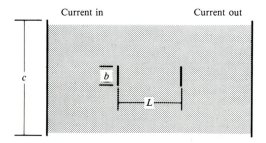

The current into the outer electrode is I.

Answer: The current that passes underneath the central electrodes is given by

$$\left(\frac{b \times W}{c \times W}\right) \times I$$

if we neglect any fringing and assume that $c \gg b$. The voltage $\Delta\phi$ that is measured across the central electrodes is given by

$$\Delta\phi = \left\{\left(\frac{b \times W}{c \times W}\right) \times I\right\} \times \left\{\frac{L}{\sigma \times (c \times W)}\right\}$$

Solving for the conductivity σ, we write

$$\frac{1}{\sigma} = \frac{\Delta\phi}{I} W\left(\frac{c^2}{bL}\right)$$

The form factor is therefore (c^2/bL) for this particular configuration.

Typical numbers for the mobility of electrons and holes for various semiconductor materials are given in Table 5.1.

The second transport coefficient that is of importance is diffusion. For this, we write that the electrons will diffuse according to

$$j_n = qD_n \frac{\partial n}{\partial x} \tag{40}$$

This has also been derived in Chapter 3.

The diffusion coefficient and the mobility are related by the Einstein relation. This can be easily derived by stating the conduction current of electrons:

$$j_n = -qnv = qn\mu_n E = -qn\mu_n \frac{\partial \phi}{\partial x} \tag{41}$$

TABLE 5.1

	μ_n	μ_p (cm^2/V-sec)
Si	1,500	600
Ge	3,900	1,900
GaAs	8,500	400
InSb	80,000	1,250
InAs	33,000	460
PbS	600	700

We will assume that the electrons obey a Maxwell–Boltzmann distribution. Therefore, (40) becomes

$$j_n = -\frac{q^2 D_n}{k_B T} n \frac{\partial \phi}{\partial x} \tag{42}$$

Equating (41) and (42), we write

$$\frac{D_n}{\mu_n} = \frac{k_B T}{q} \tag{43}$$

which is the Einstein relation.

VI. Shockley–Haynes Experiment

A classic experiment in semiconductor physics is the demonstration of the drift and the diffusion of a minority carrier in a semiconductor. This experiment was first carried out by Shockley and Haynes. The experimental setup is shown in Figure 5.9. An excess burst of minority carries $\Delta p_n = p_n - p_{n0}$ is created at $t = 0$ and at $x = 0$. An equal number burst of majority carriers is also created at this location, but we will be concerned here only with the minority carriers. Under the influence of the electric field created by the pulse generator, this burst of minority carriers will drift to the right. Its overall motion is described with the more general diffusion equation

$$\frac{\partial \Delta p_n}{\partial t} = -\mu_p E \frac{\partial \Delta p_n}{\partial x} + D_p \frac{\partial^2 \Delta p_n}{\partial x^2} - \frac{\Delta p_n}{\tau_p} \tag{44}$$

The solution of this equation follows the procedure given previously and

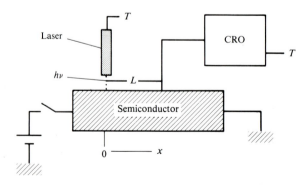

Figure 5.9 Experimental setup for the Shockley–Haynes experiment.

described in detail in Appendix A, with the additional translation in space due to the electric field understood. We find that

$$\Delta p_n = \frac{N}{\sqrt{4\pi D_p t}} \exp\left\{-\left[\frac{(x - \mu_p E t)^2}{4 D_p t} + \frac{t}{\tau_p}\right]\right\} \qquad (45)$$

This equation predicts that an initial burst will (1) decay in amplitude as $t^{-1/2}$, (2) drift with a velocity $\mu_p E$, (3) diffuse in space and time, and (4) decay in time. There are two mechanisms for the temporal change in time. The density would decay in time even if the lifetime τ_p were infinite, since we have assumed that a localized pulse containing N total particles has been created at $x = 0$ and at $t = 0$; that is,

$$N = \int_{-\infty}^{\infty} \Delta p_n \, dx = \text{constant}$$

As this fixed number of particles diffuses in space as time increases, the density must decrease in time (recall Example 3.3). The second temporal decay is due to the finite lifetime.

The measured response as obtained by Shockley and Haynes is shown in Figure 5.10 for two values of electric field, $E = 0$ and $E \neq 0$. Note that the predicted effects have occurred. For the case of $E = 0$, the particles will only recombine and diffuse. For the case of $E \neq 0$, there will be an additional drift of the particles. From these experiments, it is possible to compute the values for the various transport coefficients and the lifetime. This is done using the following procedure. The peak of the response corresponds to the term $x - \mu_p E t = 0$. Hence we can compute μ_p. The peak of the response decays in time, and the response spreads out in position as time increases. Hence we can compute D_p and τ_p from these two observations.

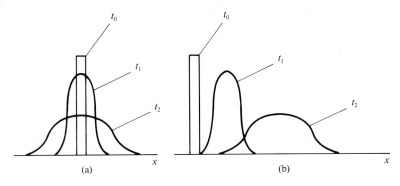

Figure 5.10 Experimental results of the Shockley–Haynes experiment. (a) $E = 0$ and (b) $E \neq 0$.

VII. Hall Effect

The most common technique to measure the carrier concentration in a semiconductor is to use the Hall effect. As an added bonus, it allows the investigator to convince others about the existence of holes as charge carriers. The Hall effect is based on the vector force that exists due to particles being in motion with a velocity **v** across a magnetic field **B**. The resulting Lorentz force is given by

$$\mathbf{F} = q(\mathbf{v} \times \mathbf{B}) \tag{46}$$

where q is the charge of the particular particle that is in motion. If we let the magnetic field be in the positive z direction and the particle velocity be in the positive x direction, as shown in Figure 5.11, then the holes will be accelerated in the negative y direction and the electrons will be directed in the positive y direction.

This separation of charged particles sets up an electric field E_y. This electric field can be computed from

$$qE_y = qv_xB_z \tag{47}$$

since the electric field balances the Lorentz force in steady state. The *Hall voltage* is given by

$$\phi_{\mathrm{H}} = E_yw = v_xB_zw \tag{48}$$

For a p-type semiconductor, we know from (37) that

$$j_p = pq\mu_pE_x = pqv_x$$

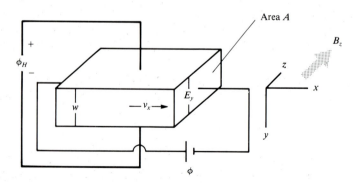

Figure 5.11   Experimental setup for the Hall-effect experiment.

The Hall electric field can be written as

$$E_y = \frac{j_p}{pq} B_z \tag{49}$$

The term

$$R_H = \frac{1}{qp} \tag{50}$$

is called the *Hall coefficient*. A similar argument is valid for the electrons, where we find that

$$R_H = -\frac{1}{qn} \tag{51}$$

Therefore, by measuring the Hall voltage ϕ_H for a known current and magnetic field, we can compute the hole density from

$$p = \frac{1}{qR_H} = \frac{j_p B_z}{qE_y} = \frac{IB_z/A}{q\phi_H/W} = \frac{IB_z W}{q\phi_H A} \tag{52}$$

All the quantities on the right side can be measured, and therefore p is determined.

VIII. Conclusion

The basic building blocks of a semiconductor device, the semiconductor materials, have now been obtained. The remaining problems are the actual manufacture and the combining of various materials into various device configurations. We will observe that the basic physics of the various devices can be understood by reading a book. The actual manufacture, however, requires experimental ingenuity and manufacturing entrepreneurship that is beyond the scope of a text. A flavor of this process is given in Chapter 9.

PROBLEMS

1. Calculate the number of electrons in the conduction band of a semiconductor $(E_c \leq E \leq E_c^*)$ at a temperature $T = 0°K$ if (a) $E_c < E_F$ and (b) $E_c > E_F$. Sketch the various densities as a function of energy E.
2. Calculate the number of holes in the valence band for the conditions given in Problem 1.
3. Show that the effective density of states given by Equation (11) is approximately equal to the density of states in a strip $1.2k_B T$ above E_c.

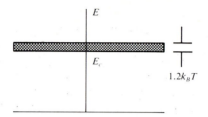

4. The effective mass of electrons and holes for pure silicon is $1.1m_e$ and $0.56m_e$, respectively, where m_e is the mass of the electron. Calculate the effective density of states for electrons and holes for silicon.

5. Derive equation (27).

6. Show that the extrinsic hole density is related to the intrinsic hole density as the temperature T increases.

7. Calculate the wavelength of light that will cause a transition from the valence to conduction bands in PbS and CdS. Are these in the visible range?

8. List possible semiconducting materials that could be used as solar cells from a scientific point of view. Investigate their applicability from an economic point of view.

9. A solar cell is used on the earth to capture the sun's energy. Show that the captured energy will increase if the cell is mounted on a platform whose normal can rotate to compensate for the earth's rotation.

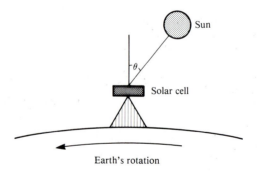

10. Find the temporal behavior of the hole density p in an intrinsic semiconductor if there is no additional generation of holes and their average velocity $\langle v \rangle$ is zero. Also, you should assume that their initial density is p_0, and the recombination rate is proportional to the density of holes times the density of electrons.

11. Calculate the drift velocity of electrons and holes for an electric field of 1 V/cm for the semiconducting materials listed in Section V. If the materials were 0.001 cm in length, estimate the frequency response of the materials due to the transit time of particles.

12. Estimate the diffusion coefficient at room temperature for electrons and holes for the materials listed in Section V.

13. Show that (45) is a solution of (44).

14. Assuming that the lifetime τ_p is infinite, obtain the solution of the general diffusion equation (44).

Chapter Six

---

pn Junctions

The creation, understanding, and manufacturing of semiconducting materials have been and continue to be a project that receives considerable attention among a certain select group of people. We could liken this challenge to the study of music. The composer creates the music in his mind, understands the meaning of the notes and scales, and then inscribes the composition that is in his mind to a piece of parchment. In both cases, the final production is understood by just a few and is hidden from the general populace until either a solid-state device is made or the music is played. The former intellectual endeavor has had a strong influence on the latter in the present-day world, probably to the dismay of some purists in both fields. The world has certainly progressed since the days when youths listened late at night to their homemade crystal receiver while their parents thought they were asleep or doing their schoolwork.

In this chapter, we will describe one of these devices. This device is the *pn* junction, and a good understanding of it is required in order to obtain a modicum of understanding or appreciation of any solid-state device. As the reader can imagine from the nomenclature, the device consists of a *p*-type semiconductor juxtaposed with an *n*-type semiconductor. We would expect that the charge carriers from one side would move over to the other and influence the characteristics of that side. Such is the case and our intuition will be rewarded with the development of several practical devices. These include diodes, rectifiers, variable capacitors, and a number of other devices. As we will see, the understanding of the operation of these devices follows a few basic principles.

I. The *pn* Junction

A *pn* junction or, rather, the joining of two semiconductors, one being of an *n* type and the other being of a *p* type, forms the basis of most semiconductor devices. Several degrees of sophistication can be employed to describe the junction, and we will initially assume the simplest model in that both sides of the semiconductor have a uniform distribution of electrons or holes at the time that the two materials are joined. This would imply a uniform doping for the extrinsic semiconductors. A typical *pn* junction just before the time of joining is shown in Figure 6.1. The energy band structure is also shown.

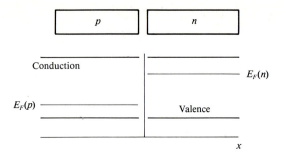

Figure 6.1 Energy level diagram of a *pn* junction just before the time of joining.

As time increases, some electrons from the *n*-type semiconductor can diffuse over the boundary and end up in the *p*-type semiconductor. There they can "annihilate" holes by recombining with them. It would be a suicide mission since the electrons are also annihilated. Therefore, in a layer just next to the junction, the *p*-type semiconductor becomes deficient in free holes. Similarly, diffusion of holes from the *p*-type semiconductor results in a layer on the *n* side becoming deficient in free electrons. If the density of these diffused particles becomes too high, they would diffuse back.

Eventually, we will achieve an equilibrium situation of having these *minority* particles (holes in an *n*-type semiconductor where the electrons would be the *majority* carrier, or electrons in a *p*-type semiconductor where the holes would be the majority carrier) reach some sort of balance. An important consequence of this free interchange of particles and hence energy is found in statistical mechanics. It states that *two systems that can exchange energy and particles are in equilibrium when the temperatures and chemical potentials are equal.* This manifests itself in causing the Fermi energies to line up at the same energy level. As shown in Figure 6.2, the energy levels of the two semiconductors have adjusted themselves such that the Fermi energies of the two materials are at the same level. This is an important conclusion.

An important consequence of this readjustment of energy levels is that an energy barrier has been set up between the two sides. The height of this energy

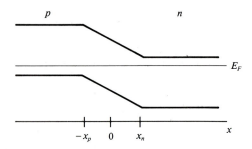

Figure 6.2 Energy level diagram of a *pn* junction after joining. The Fermi energies on either side are the same.

barrier as shown in Figure 6.1 is given by

$$q\phi_0 = E_F(n) - E_F(p) \tag{1}$$

where ϕ_0 is called the contact potential or the diffusion potential.

We would expect that these minority particles would not reside in a region of zero width but would be spread over a certain width. The width on either side can be related to the width on the other side through the following argument. In equilibrium, the total charge of positive charges Q^+ in the n-type material must equal the total negative charge Q^- that has diffused to the p-type material. Therefore,

$$Q^+ = Q^- \tag{2}$$

If there are N_d ionized donors per unit volume on the n side and N_a ionized acceptors on the p side, then (2) becomes

$$N_a x_p = N_d x_n \tag{3}$$

where x_p is the distance the space-charge region extends into the p-type side of the junction, and x_n is the distance the space-charge region extends into the n-type side of the junction. The cross-sectional area is assumed to be uniform.

II. Contact Potential

The first of the quantities that was introduced in the previous section, which described the equilibrium state that was set up when a p-type semiconductor was joined with an n-type semiconductor, was the contact potential. Since this is a thermal equilibrium situation, the total electron and hole current must separately equal zero. Recall Example 2.2.

The electron current is written as

$$j_n = 0 = qn\mu_n E + qD_n \frac{dn}{dx} \tag{4}$$

Solve for the electric field $E = -d\phi/dx$.

$$\frac{d\phi}{dx} = \frac{D_n}{n\mu_n} \frac{dn}{dx} = \frac{k_B T}{qn} \frac{dn}{dx} \tag{5}$$

where the Einstein relation has been employed. This can be integrated from $x = -x_p$ to $x = x_n$ and we obtain

$$\phi(x_n) - \phi(-x_p) = \frac{k_B T}{q} \ln\left(\frac{n(x_n)}{n(-x_p)}\right) \tag{6}$$

The contact potential ϕ_0 in equilibrium is defined from (1) as

$$\phi_0 = \phi(x_n) - \phi(-x_p)$$

It will be assumed that all the impurity atoms are ionized. Therefore, $n(x_n) = N_d$ and $p(-x_p) = N_a$, and $n(-x_p)p(-x_p) = n_i^2$. Therefore, (6) can be written as

$$\phi_0 = \frac{k_B T}{q} \ln\left(\frac{N_d N_a}{n_i^2}\right) \tag{7}$$

The dimensions of the term $k_B T/q$ are in terms of a voltage, and this is sometimes called a *thermal* potential. The voltage ϕ_0 is called the *contact* potential. Note that the n side is positive with respect to the p side.

Example 6.1 (a) If the initial dopant densities N_d and N_a are fixed at some constant value, show that the only method to increase the contact potential ϕ_0 is to increase one or both of these densities if the diode is operated at room temperature. (b) Evaluate the contact potential if the intrinsic density $n_i = 10^{10}/\text{cm}^3$ and $N_a = N_d = 10^{15}/\text{cm}^3$. Show that part (a) is true by a numerical example.

Answer: (a) From equation (7), we write

$$\Delta\phi_0 = \frac{k_B T}{q} \Delta\left\{\ln\left(\frac{N_d N_a}{n_i^2}\right)\right\} = \frac{k_B T}{q}\left(\frac{1}{N_a N_d/n_i^2}\right)\left(\frac{N_a \Delta N_d + N_d \Delta N_a}{n_i^2}\right)$$

$$= \frac{k_B T}{q}\left(\frac{\Delta N_d}{N_d} + \frac{\Delta N_a}{N_a}\right)$$

Therefore, the contact potential ϕ_0 will increase if the density of either of the dopant species is increased.
(b) The thermal potential $k_B T/q$ at room temperature is 0.026 volt. Hence

$$\phi_0 = (0.026)\ln\left(\frac{10^{15} \times 10^{15}}{10^{20}}\right) = 0.6 \text{ volt}$$

If N_a is increased to $10^{17}/\text{cm}^3$, this potential increases to 0.72 volt.

III. Depletion Width and Varactor Diodes

The second of the quantities that was introduced in Section I in the description of a *pn* junction was the concept of a region in the *p*-type semiconductor just adjacent to the junction, where all the holes had been depleted, and a region in the *n*-type semiconductor, where all the electrons had been depleted. The width of this region is called the *depletion width*, and we will find that its size depends on the doping concentration and the contact potential.

This region, where two separate parts exist, each containing a charge of a different sign, leads the reader to speculate that the junction behaves like a capacitor. The capacitance could be defined as the ratio of charge that has migrated across the junction divided by the contact potential. This speculation will be correct, as will be noted in this section. It also turns out that this capacitance will be a nonlinear function of the voltage that may be externally applied across the junction, and this has practical consequences.

The charge per unit area on the p side must equal in magnitude the charge per unit area on the n side. This manifests itself in (3), which is rewritten as

$$\int_0^{x_n} N_d(x)\, dx = \int_{-x_p}^0 N_a(x)\, dx \tag{8}$$

where the possibility of a nonuniform distribution of minority carriers is included. There may still be some majority carriers $n(x)$ and $p(x)$ on either side. The potential ϕ will satisfy Poisson's equation in two regions:

$$-x_p \le x \le 0: \qquad \frac{d^2\phi^<}{dx^2} = \frac{q[N_a(x) - p(x)]}{\varepsilon} \tag{9}$$

$$0 \le x \le x_n: \qquad \frac{d^2\phi^>}{dx^2} = \frac{q[n(x) - N_d(x)]}{\varepsilon} \tag{10}$$

where the superscripts refer to the two separate regions and ε is the dielectric constant of the material.

It is possible to simplify (9) and (10) by first assuming that the densities of free carriers $n(x)$ and $p(x)$ are much less than the densities of impurity atoms on each side. The second approximation that must be made is to describe the density variation of impurity atoms as functions of position x. The simplest approximation is to assume that this charge per unit volume is constant in the two regions. This is shown in Figure 6.3.

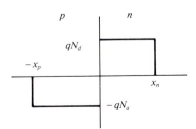

Figure 6.3   Charge distribution at the edge of a *pn* junction.

We have just to solve the equations

$$-x_p \leq x \leq 0: \qquad \frac{d^2\phi^<}{dx^2} = \frac{qN_a}{\varepsilon} \tag{11}$$

$$0 \leq x \leq x_n: \qquad \frac{d^2\phi^>}{dx^2} = -\frac{qN_d}{\varepsilon} \tag{12}$$

The integral of (11) is

$$\phi^< = \frac{qN_a x^2}{2\varepsilon} + Ax + B$$

and the integral of (12) is

$$\phi^> = -\frac{qN_d x^2}{2\varepsilon} + Cx + D$$

where A, B, C, and D are constants of integration. They are evaluated from the following boundary conditions. The contact potential is defined from

$$\phi_0 = \phi(x_n) - \phi(-x_p)$$
$$= \phi^>(x_n)$$

where we have arbitrarily set $\phi(-x_p) = \phi^<(-x_p) = 0$. Therefore,

$$0 = \frac{qN_a x_p^2}{2\varepsilon} - Ax_p + B \tag{i}$$

$$\phi_0 = -\frac{qN_d x_n^2}{2\varepsilon} + Cx_n + D \tag{ii}$$

The potential should be continuous at $x = 0$. Therefore,

$$B = D \tag{iii}$$

Also, the electric fields at $x = -x_p$ and at $x = x_n$ are equal to zero since (2) is satisfied, and Gauss's law would yield 0. Therefore,

$$0 = \frac{qN_a x_p}{\varepsilon} - A \tag{iv}$$

$$0 = \frac{qN_d x_n}{\varepsilon} - C \tag{v}$$

A final boundary condition is that the electric field must be continuous at $x = 0$. Therefore,

$$A = C \tag{vi}$$

The simultaneous solution of these six equations (i–vi) in six unknowns, A, B, C, D, x_n, and x_p, leads to

$$x_n = \left\{ \frac{2\varepsilon\phi_0}{q} \left[\frac{N_a}{N_d(N_a + N_d)} \right] \right\}^{1/2} \tag{13}$$

$$x_p = \left\{ \frac{2\varepsilon\phi_0}{q} \left[\frac{N_d}{N_a(N_a + N_d)} \right] \right\}^{1/2} \tag{14}$$

The total *depletion width* is given by

$$W = x_n + x_p = \left\{ \frac{2\varepsilon\phi_0}{q} \left[\frac{N_a + N_d}{N_d N_a} \right] \right\}^{1/2} \tag{15}$$

This is the width of the depletion layer. Other approximations could be made for the distributions of the charge densities in the two regions, but the procedure would be the same. An important distribution would be a linear variation of density within the depletion layer.

We note on inspection of (15) that the depletion width depends on the barrier potential ϕ_0; in fact, it is not a linear dependence, but is nonlinear. Let us suggest a slight modification to the system, that of applying a reverse-bias potential across the junction. Reverse bias is defined as making the p side negative with respect to the n side. The convention that will be employed is that if a forward-bias potential is positive then the contact potential ϕ_0 given in (7), which is just a function of the doping concentrations and the temperature, should be replaced with $(\phi_0 - \phi_a)$, where ϕ_a is the potential that is applied across the junction. Equation (15) then becomes

$$W = \left\{ \frac{2\varepsilon(\phi_0 - \phi_a)}{q} \left[\frac{N_a + N_d}{N_d N_a} \right] \right\}^{1/2} \tag{16}$$

This is an excellent approximation for values of ϕ_a that are large and negative. It is not valid for forward-bias conditions since there would then be an increased number of carriers in the space-charge region.

Example 6.2   Calculate the total depletion width W and the depletion widths in each region of a pn junction at room temperature. The dopant densities are $N_a = 10^{17}/\mathrm{cm}^3$, $N_d = 10^{15}/\mathrm{cm}^3$, and $n_i = 10^{10}/\mathrm{cm}^3$. The relative dielectric constant $\varepsilon_s = 12$. Be careful with units.

Answer: The contact potential is calculated by Equation (7) to be $\phi_0 = 0.72$ volt. From Equation (15), we write

$$W = \left\{ \frac{(2)(12)[(1/36\pi) \times 10^{-9}](0.72)}{1.6 \times 10^{-19}} \times \frac{10^{23} + 10^{21}}{10^{23} \times 10^{21}} \right\}^{1/2}$$

$$= 0.98 \times 10^{-4} \text{ cm} = 0.98 \ \mu\text{m}$$

From charge conservation stated by equation (3), we have $N_a x_p = N_d x_n$. Also, $W = x_n + x_p = x_n[1 + (N_d/N_a)] = x_n[1 + 0.01]$. Hence $x_n = 0.97 \ \mu\text{m}$ and $x_p = 0.01 \ \mu\text{m}$. Since the p side is doped greater than the n side, $x_n > x_p$. The junction depletes further into the lighter doped material.

This depletion width has some important practical implications in that a reverse-biased *pn* junction can be the basis for a fundamental electrical device, the *varactor* or *varicap*. From (16), we find that the depletion width increases as the applied potential ϕ_a is made more negative. This change in the depletion width results in variations in the immobile charge that is uncovered on the two sides of the junction. The capacitance per unit area of this junction is defined from

$$C = \left| \frac{dQ}{d\phi_a} \right| \tag{17}$$

where Q is the charge per unit area on one side of the junction. On the p side, this has the value of $Q = qN_a x_p$. Using (14), we write (17) as

$$C = \left| \frac{d\left\{ qN_a \left[\dfrac{2\varepsilon(\phi_0 - \phi_a)}{q} \dfrac{N_d}{N_a(N_a + N_d)} \right]^{1/2} \right\}}{d\phi_a} \right| = \left\{ \left| \frac{\varepsilon q N_d N_a}{2(\phi_0 - \phi_a)(N_a + N_d)} \right| \right\}^{1/2}$$

This can be written as

$$C = \frac{C_0}{\sqrt{1 - (\phi_a/\phi_0)}} \tag{18}$$

where

$$C_0 = \sqrt{\frac{\varepsilon q N_d N_a}{2\phi_0(N_a + N_d)}}$$

which would be the capacitance of the "parallel-plate capacitance" of the unbiased depletion width. A voltage-dependent capacitor has many practical applications, such as tuning FM receivers. This analysis does not work under forward-bias conditions, which will be examined below.

If the shunt elements in an *LC* transmission line were replaced with these elements, we would construct a nonlinear transmission line that could admit the propagation of solitons, nonlinear waves that were first observed as waves in water and later observed in plasmas. The electrical signal carried along the nerve and the red spot on Jupiter have soliton properties, and it is a topic of much current research. This will be further described in Chapter 11.

IV. *pn* Diode Characteristics

In the previous section, the static characteristics of a *pn* junction were described. It was found that under the conditions of reverse bias (*p* side negative with respect to the *n* side) a negligible current would flow. The small current that does flow is called the *reverse current*. However, if the polarity of the bias is reversed, there can be a large flow of current from the *p* region to the *n* region since we would be transporting majority carriers across the boundary from a large storehouse (i.e., holes from the *p* side and electrons from the *n* side). This difference in flow capabilities suggests a possible application of a *pn* junction, that of acting as a *rectifier*. If a *pn* diode is placed in series with a resistor and an ac voltage source, the voltage as detected across the resistor will be of only one polarity (if we neglect the small reverse current).

Prior to using any mathematics to describe the operation of a *pn* diode as a circuit element, we will present a qualitative description. The understanding gained in the previous section should aid the reader when we later introduce the mathematical description. For the moment, we will assume that an ideal *pn* diode is being analyzed. This diode has no electrical resistance, and any applied voltage bias appears only at the actual junction even if a current is flowing.

Consider the series of three *pn* junctions with three values of bias applied across the junction as shown in Figure 6.4. In addition to the diodes, the potential distribution and the currents of each type are listed. These currents are separated into four components, electron and hole diffusion and conduction.

For the case of zero applied voltage bias, the electron diffusion current balances the electron conduction current, resulting in no net electron current. A similar description applies to the hole current. The diffusion current consists of majority carrier electrons on the *n* side that have enough energy to climb the potential barrier and diffuse into the *p* side. It also consists of majority carrier holes on the *p* side that have migrated to the *n* side. The conduction current consists of the flow of the minority particles across the junction. Since the density of these particles is small, we would expect the current created by this component to be small and to be relatively unaffected by changes in the bias. The voltage that appears across the junction is the contact potential ϕ_0. A depletion width that is proportional to $\sqrt{\phi_0}$ is obtained if the distribution of minority carriers is uniform. This is shown schematically in Figure 6.4(a).

If a reverse bias equal to $-\phi_R$ is applied across the junction, the depletion width will increase since the width is proportional to $(\phi_0 + \phi_R)^{1/2}$. The electron and hole diffusion currents decrease, and the conduction currents will remain unaffected. This is shown in Figure 6.4(b).

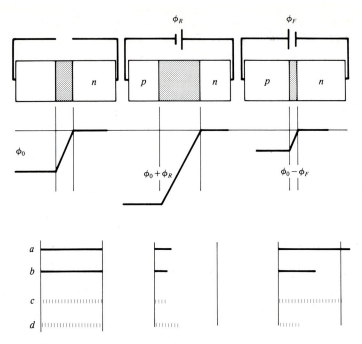

Figure 6.4 The currents are (a) hole diffusion from *p* to *n*, (b) hole conduction from *n* to *p*, (c) electron diffusion from *p* to *n*, and (d) electron conduction from *n* to *p*.

If now the polarity of the bias is changed to a value that will be called a forward bias ϕ_f, the depletion width will be decreased. [See Figure 6.4(c).] The majority carriers will find a lower potential barrier and can more easily move across this region and enter the other side. Therefore, we would expect and do indeed obtain a larger net current, since the dominant carriers that carry the current are the majority carriers rather than the minority carriers, as in the case of reverse bias.

It will be shown that the current through a diode can be written as a function of the potential applied across the diode. This equation is called the diode equation and is written as

$$I = I_0[e^{q\phi/(k_B T)} - 1] \tag{19}$$

where I_0 is the value of the current under conditions of a very large reverse bias. It is called the *reverse saturation current*. At a voltage $\phi = 0$, the current $I = 0$. The current exponentially increases as ϕ is made more positive in the forward-bias region. A sketch of the I–ϕ characteristics of the diode equation (19) is shown in Figure 6.5. We will now derive (19).

To derive the diode equation, we must inquire what happens to the excess minority carriers that reside within the depletion region. The excess carriers may

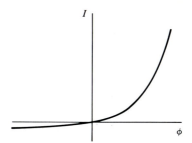

Figure 6.5 Voltage–current
characteristics of a *pn* junction diode.

have been created due to the externally applied voltage or possibly due to an
incident photon of light. The additional contribution will be defined as

$$\Delta n_p = n_p - n_{p0} \tag{20}$$

where n_{p0} is the equilibrium density of minority carriers, in this case electrons on
the *p* side of the junction. A similar treatment follows for holes on the *n* side. The
temporal and spatial evolution of these minority carriers is governed by the
diffusion equation

$$\frac{\partial \Delta n_p}{\partial t} - D_e \frac{\partial^2 \Delta n_p}{\partial \beta^2} = -\frac{\Delta n_p}{\tau_e} \tag{21}$$

In (21), D_e is the diffusion coefficient for electrons, τ_e corresponds to a
recombination lifetime, and the spatial variable $\beta = x - x_p$. It will be assumed
that these minority electrons will only be absorbed on the *p* side and none will
migrate back to the *n* side. See Figure 6.6.

In the steady state, (21) can be written as

$$D_e \frac{\partial^2 \Delta n_p}{\partial \beta^2} = \frac{\Delta n_p}{\tau_e} \tag{22}$$

The solution of (22) is

$$\Delta n_p(\beta) = Ae^{-\beta/L_e} + Be^{\beta/L_e} \tag{23}$$

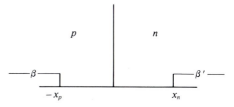

Figure 6.6 Coordinate systems for
studying the *pn* junction diode. To
avoid any minus signs, it is
convenient to define two coordinate
systems.

where $L_e = (D_e \tau_e)^{1/2}$. The boundary conditions that are applicable are that as $\beta \rightarrow \infty$ the minority density contribution should go to zero due to its recombination with the majority particles. Therefore, the constant of integration $B = 0$. The second boundary condition states that at the edge of the depletion layer the minority distribution should follow a Maxwell–Boltzmann distribution.

$$n_p(x = x_p) = n_p(\beta = 0) = n_{p0}e^{q\phi/(k_BT)} \tag{24}$$

This determines the constant A, and we finally write for the region $\beta > 0$

$$\Delta n_p(\beta) = n_{p0}[e^{q\phi/(k_BT)} - 1]e^{-\beta/L_e} \tag{25}$$

This excess charge will diffuse into the p-type semiconductor, and it is possible to calculate the flux associated with this diffusion from

$$j_{ep} = -qD_e \frac{\partial \Delta n_p}{\partial \beta}$$

We find that

$$j_{ep} = \frac{qD_e n_{p0}}{L_e}[e^{q\phi/(k_BT)} - 1]e^{-\beta/L_e} \tag{26}$$

In a similar fashion, the flux of minority holes in the n side is computed to be

$$j_{hn} = -\frac{qD_h p_{n0}}{L_h}[e^{q\phi/(k_BT)} - 1]e^{\beta'/L_h} \tag{27}$$

where $\beta' = x - x_n$ and $L_h = (D_h \tau_h)^{1/2}$. The total current is therefore the sum of (26) and (27) evaluated at the edges of their respective depletion layers, $\beta = 0$ and $\beta' = 0$, respectively. We write

$$I = A[j_{ep}(\beta = 0) + j_{hn}(\beta' = 0)]$$

where A is the cross-sectional area of the diode. Therefore, we write finally

$$I = I_0[e^{q\phi/(k_BT)} - 1] \tag{19}$$

where

$$I_0 = -qA\left[\frac{D_h p_{n0}}{L_h} + \frac{D_e n_{p0}}{L_e}\right] \tag{28}$$

If we had started with the model that the diode was not infinite in length but rather finite, the term I_0 would have to be modified to incorporate this change. This current I_0 is called the *reverse saturation current*.

Example 6.3   A value for the reverse saturation current I_0 is found to be $I_0 = 1\ \mu A$ at room temperature. Find the voltages that will produce a forward current of 1 mA and 1 A.

Answer: From Equation (19), we write

$$10^{-3} = 10^{-6}[e^{\phi/0.026} - 1]$$

$$\phi = (0.026)\ln(1001) = 0.18 \text{ volt}$$

Also,

$$\phi = (0.026)\ln(1{,}000{,}001) = 0.36 \text{ volt}$$

In this derivation, we have considered only the behavior of the minority carriers when calculating the current. The majority carriers also are important and present in the diode. Since we are considering only a one-dimensional system, the total current must remain a constant. The diffusion of excess minority carriers must be replaced by the diffusion of excess majority carriers. The diffusion of the majority carriers will be in the opposite direction as the minority carriers. However, the charge is of the opposite sign, so the current will be in the same direction. Recall that the majority carrier conduction current may be larger than the majority carrier diffusion current and is in the opposite direction. The minority carrier conduction current was shown to be negligible. The components of the current of a forward-biased *pn* junction are shown in Figure 6.7.

V. Reverse Breakdown

According to (19), the current will saturate as the bias voltage is made sufficiently negative. If there were no other physical principles involved, this current could be easily calculated and this region of operation would be a rather dull region. Fortunately, two additional physical phenomena should be considered that will lead to practical devices. The devices are based on two breakdown phenomena, *avalanching* (Zener breakdown) and *tunneling*.

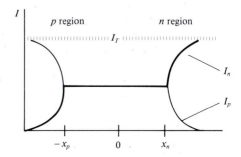

Figure 6.7   Components of the total current I_T of a forward-biased *pn* junction.

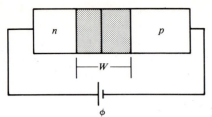

Figure 6.8   Reverse-biased *pn* junction.

Consider a diode that has a large reverse bias applied to it as shown in Figure 6.8. A minority electron will be accelerated by the electric field $E = \phi/(x_n + |x_p|) = \phi/w$. If this electron has gained sufficient energy, it could create a hole–electron pair upon collision with an electron in the valence band. These new electron–hole pairs could be also accelerated due to the electric field, creating further electron–hole pairs. The process has essentially avalanched, and diodes based on this process are called avalanche diodes. There is some external control in the manufacturing process, since the depletion width depends on the density of the minority carriers. Also, the voltage dependence of the maximum electric field that can be applied to the diode to accelerate the initial electrons can be estimated as

$$E\big|_{max} \approx \frac{\phi}{W} \approx \frac{\phi}{K\sqrt{\phi}} = \frac{\sqrt{\phi}}{K}$$

where K is a constant that includes the doping densities and a uniform distribution of minority carriers is assumed.

The second mechanism for the creation of a reverse-biased diode is based on a quantum phenomenon. Let the reverse-bias potential be sufficiently large such that the energy level of the valence band of the *p*-type semiconductor is at approximately the same level as the conduction band of the *n*-type semiconductor, as shown in Figure 6.9. There is then a nonzero probability that an electron from the valence band will "tunnel" across to the conduction band, which will create new majority carriers. From (16), it is noted that the depletion

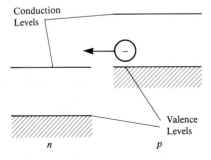

Figure 6.9   Highly reverse biased *pn* junction. The electron "tunnels" from the valence band of the *p* region to the conduction band of the *n* region.

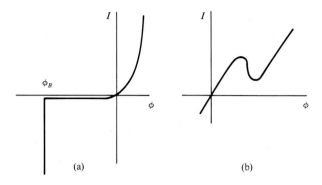

Figure 6.10   Voltage–current characteristics of (a) an avalanche and (b) a tunnel diode.

width w will decrease as the density of the impurities at the junction is increased. This increases the probability of this tunneling actually taking place.

The tunneling and the avalanching mechanisms can also be separated in an experimental situation. If the reverse voltage is small, we usually encounter the tunneling mechanism. If it is larger such that an ionization can occur, then the dominant mechanism is the avalanching mechanism. The voltage–current characteristics of these diodes are shown in Figure 6.10.

We can think of a practical application for a reverse-biased diode, that of a voltage regulator. The resistance of the diode switches from a value of $R \to \infty$ to a value of $R \to 0$ at a critical voltage. The voltage at the output of a device would stay constant if a tunnel or avalanche diode were connected across the output terminals. A practical application for a tunnel diode since it possesses a region of negative resistance $(d\phi/dI < 0)$ is that it could be used as an amplifier. The application of a tunnel diode as an amplifier is not widely used now, although the tunneling mechanism is used in a scanning tunnel electron microscope to probe the mysteries of atomic structure.

VI. Switching Characteristics

The knowledge of the temporal switching characteristics from forward to reverse bias or the opposite switching of a diode is important since these characteristics govern the high-frequency behavior of the diode (i.e., they determine the bandwidth for operation). Since the depletion width is a function of the minority carrier density that resides within it and we have some control of this minority density through doping, it is possible to create a "one-sided" diode in which one side is more heavily doped than the other. Let us therefore consider a one-sided diode with the p side more heavily doped than the n side. The charge stored in the n side can then be considered to be in the form of the distribution of the excess hole concentration.

The temporal and spatial evolution of this hole concentration is governed by the continuity equation for excess holes:

$$-\frac{\partial j_p}{\partial x} = \frac{qp}{\tau_p} + q\frac{\partial p}{\partial t} \tag{29}$$

This equation can be integrated from the edge of the depletion layer x_n to the edge of the semiconductor, which we will take to be at $x \to \infty$. We find that

$$j_p(x_n) - j_p(\infty) = q\int_{x_n}^{\infty}\left(\frac{p}{\tau_p} + \frac{\partial p}{\partial t}\right)dx \approx j_p(x_n) \tag{30}$$

where the current density at $x \to \infty$ (or at the end of a long semiconductor) is taken to be zero.

The integral can be interpreted as being the total charge per area A that is enclosed within the depletion layer. Hence the current i_p is given by

$$i_p(x_n, t) = \frac{Q_p(t)}{\tau_p} + \frac{\partial Q_p(t)}{\partial t} \tag{31}$$

In the steady state, this has the value of $Q_p = I\tau_p$. To solve (31), we must make an assumption about the state of the diode before the switching takes place. If it were forward biased originally before the switching to a reverse bais took place at $t = 0$, then a steady-state forward current I_f would exist for $t < 0$. At $t = 0^+$, the current would instantaneously be $-I_r$.

With these switching conditions, we can obtain the solution of (31) using a Laplace transform and find that

$$Q_p(t) = I_f\tau_p e^{-t/\tau_p} + I_r\tau_p[e^{-t/\tau_p} - 1] \tag{32}$$

A switching time, t_s, can be defined as the time when this charge $Q_p(t = t_s)$ equals zero. This is found from (32) as

$$I_r = (I_f + I_r)e^{-t_s/\tau_p}$$

or

$$t_s = \tau_p\ln\left(1 + \frac{I_f}{I_r}\right) \tag{33}$$

After a time governed by this switching time, the diode can start to support a reverse voltage since the remainder of the excess holes is removed from the n region with a time constant equal to the excess-carrier lifetime. In Figure 6.11, the temporal current and voltage characteristics of a diode that is switched from forward to reverse bias are shown. This switching can be accomplished in a shorter time using a metal–semiconductor diode.

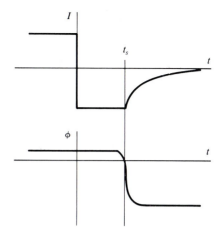

Figure 6.11 Temporal characteristics of a diode switched from forward to reverse bias.

VII. Metal–Semiconductor Diodes

In the junctions described in the previous sections, an *n*-type and a *p*-type semiconductor were joined together. An equivalent and possibly somewhat simpler junction can be formed if a semiconductor is joined together with a metal. Many of the useful properties of the *pn* junction can be found in the metal–semiconductor junction; in particular, it can be used in high-frequency applications.

To understand the operation of a metal–semiconductor diode, consider the energy band structure of the metal and the *n*-type semiconductor shown in Figure 6.12. In Figure 6.12(a), the band structure is shown prior to the joining, and in Figure 6.12(b), the equilibrium structure is shown after the joining. In the latter case, since an equilibrium system is modeled, the Fermi energies have equilibrated to the same level.

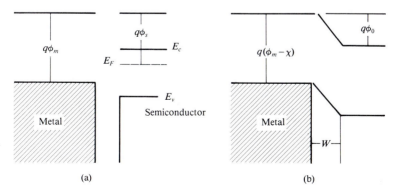

Figure 6.12 Band structure (a) before and (b) after joining a metal–semiconductor diode.

The metal will have a *work function* $q\phi_m$, where the work function of a metal is defined as the energy required to remove an electron at the Fermi level of the metal to the outside vacuum region. For very clean aluminum or gold surfaces, this energy is 4.3 and 4.8 eV, respectively. If this electron resides near the metallic surface, it will induce a positive image charge in the metal. This will tend to reduce the effective work function of the metal vacuum interface if we include the applied electric field, which is called the *Schottky effect*. The generic class of metal–semiconductor devices is called the *Schottky barrier diodes*, although other effects enter into the complete description of these devices.

In the process of aligning the Fermi energies, the electrostatic potential of the semiconductor must be altered with respect to the metal. If we join an *n*-type semiconductor to a metal as shown in Figure 6.12, this realignment creates a depletion layer in the semiconductor. In the case of the metal–semiconductor diode, the uncompensated donor ions within the depletion layer balance the negative charge on the metal. This is similar to the depletion layer that was found in a *pn* diode.

The equilibrium contact potential ϕ_0, which will inhibit further electron diffusion from the semiconductor to the metal, is just the difference of the work functions $\phi_m - \phi_s$ of the metal and the semiconductor. This value can be either increased or decreased with the application of a reverse-bias or forward-bias voltage, as was found in *pn* diodes.

The current–voltage characteristics of a metal-semiconductor diode are governed by the diode equation (19), which is rewritten as

$$I = I_0[e^{q\phi/(k_B T)} - 1] \tag{19}$$

The reverse saturation current is more difficult to derive in this case, although we would intuitively expect that it would depend on the height of the barrier that prevents electrons from being injected from the metal into the semiconductor.

The diode can act as a rectifier in that there is easy current flow from the semiconductor to the metal when the diode is based in the forward direction and very little current if the diode is biased in the reverse direction. The forward current in the case of an *n*-type metal or a *p*-type metal is due to the injection of *majority* carriers from the semiconductor into the metal. This class of diodes is sometimes called majority carrier devices. The high-frequency properties and switching speeds are better than for typical *pn* junctions.

An astute reader is probably wondering at this point whether we always obtain an effective metal–semiconductor junction every time a wire or a contact is made in a circuit? The answer to this question is a qualified yes. The qualifier is that, if the charge induced in the semiconductor in order to align the Fermi levels of the two materials is provided by the majority carriers, then this rectifying effect will be small. Hence the junction will be *ohmic* in that there will be a straight-line relation between the current and voltage. In practice, we can heavily dope the region where the contact is to be made to produce an ohmic or low-resistance contact.

VIII. Equivalent Circuit

The equivalent circuit for an ideal diode is easy to draw. For potentials less than zero (reverse bias), the current that can flow is zero for any value of applied potential. This would correspond to an open circuit. For potentials greater than zero (forward bias), any value of current can flow with no change in potential (i.e., a short circuit). That was simple in an ideal world. However, the world is not ideal and a more realistic model for the diode must be formulated.

As a first approximation, the diode has a finite conductance g that can be computed from the diode equation (19):

$$I = I_0[e^{q\phi/(k_BT)} - 1] \tag{19}$$

This is found from the definition of the conductance

$$g = \frac{dI}{d\phi} \tag{34}$$

from which we compute

$$g = I_0\left(\frac{q}{k_BT}\right)e^{q\phi/(k_BT)} \tag{35}$$

Combining (35) and (19), we can write

$$g = \left(\frac{q}{k_BT}\right)(I + I_0) \tag{36}$$

where I is the total diode current. The incremental resistor r is just

$$r = \frac{k_BT}{q(I + I_0)} \tag{37}$$

The values given by (36) and (37) are valid for small values of forward and reverse currents. If the values computed from these two equations, which govern the values at the junctions, become comparable with similar quantities of the bulk semiconductor materials or due to leakage effects, then the model has to be improved.

The described effect constitutes a static characteristic of the diode. Due to the wide application of semiconductor diodes in time-varying situations, we would expect that the model should also incorporate other circuit elements. Such is the case. The first additional element that must be incorporated into the model is a capacitance that reflects the charge separation in the depletion layer. This *depletion layer capacitance* was discussed in Section III. The second additional element arises from the consideration of the behavior of excess minority carriers stored in the neutral regions under conditions of forward bias and is called the storage or diffusion capacitance.

This *diffusion capacitance* may be larger in magnitude than the depletion-layer capacitance for moderate values of forward bias. It will first be described qualitatively as follows. The current due to the minority electrons in the *p* region is due to diffusion created by the gradient in the excess carriers whose density is Δn_p:

$$j_{ep} = qD_e \frac{d\,\Delta n_p}{d\beta} \tag{38}$$

where $\beta = 0$ at the edge of the space-charge layer. From (25), the spatial dependence of this minority charge is given by

$$\Delta n_p(\beta) = n_{p0}[e^{q\phi/(k_B T)} - 1]e^{-\beta/L_e} \tag{25}$$

This assumes that the result calculated under static conditions is valid under dynamic conditions. It is reasonable if the junction voltage does not change on a time scale on the order of the lifetime of the minority carriers. With this approximation, we note that the minority charge depends on the junction voltage. The charge and the voltage can change only at the same rate. The current (38) depends on the gradient of the excess density and thus the current can change more readily than the junction voltage.

The diffusion capacitance can be analyzed in a manner that is similar to the treatment for the depletion-layer capacitance. The total charge of electrons in the *p* region is given by

$$Q_{ep} = -\int_0^\infty qA\,\Delta n_p(\beta)\,d\beta \tag{39}$$

This integral can be performed if (25) is used in (39) and we find

$$Q_{ep} = -qA n_{p0} L_e [e^{q\phi/(k_B T)} - 1] \tag{40}$$

The capacitance associated with this stored charge is found from

$$C_e = -\frac{dQ_e}{d\phi} = \left[\frac{q^2 A n_{p0} L_e}{k_B T}\right] e^{q\phi/(k_B T)} \tag{41}$$

A similar expression is found for the holes in the *n*-type semiconductor, and the total diffusion capacitance is the sum of the two.

$$C_{\text{diffusion}} = \left[\frac{q^2 A}{k_B T}\right][p_{n0}L_h + n_{p0}L_e]e^{q\phi/(k_B T)} \tag{42}$$

There is an alternative way of writing (42) if we incorporate the definitions $L_e = (D_e\tau_e)^{1/2}$ and $L_h = (D_h\tau_h)^{1/2}$ with (28). The reverse saturation current becomes

$$I_0 = -qA\left[\frac{L_h p_{n0}}{\tau_h} + \frac{L_e n_{p0}}{\tau_e}\right] \tag{43}$$

Therefore, if the minority carrier lifetimes are approximately equal, $\tau_e \approx \tau_h = \tau$, (42) can be written as

$$C_{\text{diffusion}} = -\frac{q\tau}{k_B T} I_0 e^{q\phi/(k_B T)} \tag{44}$$

For the case of large forward bias, this can be written as

$$C_{\text{diffusion}} = -\frac{q\tau}{k_B T} I \tag{45}$$

where the diode equation (19) has been employed. Writing the diffusion capacitance in this form highlights the fact that the diode current has the same functional dependence on the junction voltage as the diffusion capacitance.

It is now possible to suggest a small-amplitude equivalent circuit for the junction diode as shown in Figure 6.13. The resistor r is defined by (37) and the resistor R_b is the resistance of the bulk of the semiconductor. The capacitor C_{junction} represents the sum of the depletion layer and the diffusion capacitances. To this simple model, we could add the inductance of the connecting wires and the capacitance between these connectors, which may be important at high frequencies.

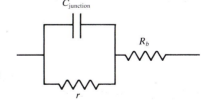

Figure 6.13 Equivalent circuit of *pn* junction.

Example 6.4   Estimate the high-frequency cutoff frequency for a *pn* diode.

Answer: The high-frequency cutoff frequency can be estimated by equating the real part of the impedance of the junction with the bulk resistance of the diode. From Figure 6.13, we write

$$R_b = \frac{r/(i\omega C_{\text{junction}})}{r + [1/(i\omega C_{\text{junction}})]}$$

or

$$\omega^2 \approx \frac{1}{r R_b (C_{\text{junction}})^2}, \qquad \text{if } r \gg R_b$$

IX. Optical Diodes

The absorption of optical energy by semiconducting diodes may create hole–electron pairs that would contribute to the electrical properties of the material. Similarly, the recombination of hole–electron pairs may result in the emission of optical photons. The fundamental concepts of absorption and radiation were introduced in Chapter 1, and in this section we will describe the application of these effects in *pn* junction devices.

A *photodiode* is a reverse-biased *pn* junction designed to employ optical generation in and near the junction and is used as a light detector. The current in a "dark" photodiode is the reverse-bias saturation current I_0. This saturation current is due to the minority carriers that are thermally generated within a few diffusion lengths on either side of the space-charge layer. If light of the proper frequency is incident on the junction, both minority and majority carriers are produced in approximately equal numbers. The percentage of change of minority carriers will be larger than that of the majority carriers and, in fact, their number may exceed the number of thermally generated minority carriers. Therefore, the reverse saturation current in this case will be determined by optical generation.

In addition to just detecting the presence of incident light, the absorption of light can be used to generate electric power. This device is called a *solar cell*. If a hole–electron pair is created in an unbiased *pn* junction, the built-in field of the contact potential drives each of the charges to the side of the junction where it is a majority carrier. The motion of these carriers constitutes a reverse current across the diode and it will produce a detectable drop in the voltage. This reduced barrier potential will increase the normal forward current flow of the *pn* diode. When operated as a photocell or a solar cell, the optically generated reverse current is larger in magnitude than the component of current that flows in response to the forward bias, so voltage drop across the diode is positive while the current is negative.

This current can be written as

$$I = I_L + I_0[1 - e^{q\phi/(k_BT)}] \tag{46}$$

where I_L corresponds to the current due to the light and the second term represents the reverse diode current. The calculation of I_L is straightforward in that we need only find the minority carrier density distributions that are generated by the light and hence their current due to diffusion.

The minority density is calculated from the steady-state diffusion equation, which includes both the effects of optical generation G_L and recombination.

$$D_p \frac{\partial^2 \Delta p_n}{\partial x^2} + G_L - \frac{\Delta p_n}{\tau_p} = 0 \tag{47}$$

As before, a length is defined as $L_p = [D_p\tau_p]^{1/2}$ for this minority hole density in the n-type semiconductor. If the generation rate G_L is spatially homogeneous, then the solution of (47) that is valid for the n-type semiconductor occupying the space $x > 0$ is given by

$$\Delta p_n = G_L\tau_p + Ae^{-x/L_p}$$

where A is a constant of integration. At $x = 0$, this constant can be determined since the cell acts as a normal diode, and we have

$$\Delta p_n|_{x=0} = p_{n0}[e^{q\phi/(k_BT)} - 1]$$

and hence A can be specified. We finally obtain

$$\Delta p_n = \frac{G_L L_p^2}{D_p}[1 - e^{-x/L_p}] + \{p_{n0}[e^{q\phi/(k_BT)} - 1]\}e^{-x/L_p} \tag{48}$$

In a similar manner, we derive that

$$\Delta n_p = \frac{G_L L_n^2}{D_n}[1 - e^{-x/L_n}] + \{n_{p0}[e^{q\phi/(k_BT)} - 1]\}e^{-x/L_n} \tag{49}$$

The hole current of the cell is computed from

$$I_p(x) = -qAD_p \frac{d\Delta p_n}{dx}$$

and the electron current is computed from

$$I_n(-x) = qAD_n \frac{d\,\Delta n_p}{dx}$$

The total generated current created by the incident light is just the sum of these two terms evaluated at $x = 0$ and evaluated at the potential of $\phi = 0$. From this, we compute that

$$I_L = qG_L(L_n + L_p)A \qquad \qquad \textbf{(50)}$$

Combining (50) with (46), we finally obtain

$$I = I_L + I_0[1 - e^{q\phi/(k_BT)}] \qquad \qquad \textbf{(51)}$$

The power that can be generated from a solar cell is given by the product of the current given by (51) times the voltage ϕ. Before actually carrying out the details of finding the maximum value of this power, it is convenient to examine the current–voltage characteristics, which are shown in Figure 6.14(a). It will be assumed that the diode is connected to a resistor as shown in Figure 6.14(b). The attempt to find the power delivered by the diode to the resistor is equivalent to finding the area of the rectangle whose opposite corners are at $(0,0)$ and (ϕ, I), where I is computed from (51). The maximum power is just the rectangle with the largest area that can fit under the curve.

If we carry out the details of computing the voltage $\phi = \phi_{mp}$, we find that a transcendental equation,

$$I_L + I_0 = I_0\left[1 + \frac{\phi_{mp}}{(k_BT)/q}\right]e^{q\phi_{mp}/(k_BT)}$$

must be solved. This is found by setting $dP/d\phi = d(I\phi)/d\phi = 0$. The efficiency η

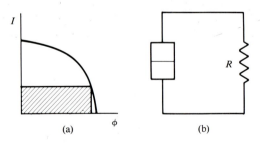

(a)   (b)

Figure 6.14   (a) Voltage–current characteristics of a solar cell. The area of the rectangle is proportional to the output power. (b) Circuit application of a solar cell.

is computed from

$$\eta = \frac{I_{mp}\phi_{mp}}{P_{in}}$$

and has a value of 10% to 25%.

Example 6.5   One can approximate the current in an optical diode as

$$I = I_0\left[1 - \left(\frac{\phi}{\phi_0}\right)^2\right]$$

Find the value of voltage that causes maximum power operation.

Answer: The power is defined from $P = I \times \phi$ or

$$P = I_0\left[1 - \left(\frac{\phi}{\phi_0}\right)^2\right]\phi$$

Find

$$\frac{dP}{d\phi} = 0 = I_0\left[1 - \left(\frac{\phi}{\phi_0}\right)^2\right] - 2I_0\left(\frac{\phi}{\phi_0}\right)^2$$

or $\phi_{max\,power} = \phi_0/(3)^{1/2}$. The maximum current is $2I_0/3$ and the corresponding power is $2I_0\phi_0/3(3)^{1/2}$.

If a forward bias is applied across a *pn* junction, the carriers are injected across the junction to establish excess carriers above their thermal equilibrium values. These excess carriers recombine and energy is released in terms of light. This generic form of diode is called a *light-emitting diode* (LED). The wavelength of the radiated energy should be in the visible range to be effective. This is usually obtained using materials that have a direct rather than an indirect transition between energy bands.

More complicated assumptions for the generation mechanism or the assumption of a finite size of the diode can be used, but they just lead to further complications in the mathematics. A more interesting question that must be approached is how to make such a device useful. The criterion that is most often applied is that the light should be in the visible range for either a solar cell or for a light-emitting diode. This criterion specifies that the gap energy, which is given by

$$E_g = hv \tag{52}$$

must have a value of 1.7 eV $< E_g <$ 3 eV. Various semiconducting compounds can be used to fulfill this requirement.

X. Conclusion

The *pn* junction diode has found wide application in various electrical circuits. It can be and has been used as a protective circuit element, a nonlinear circuit element, a "one-way" element, and an element that allows the reader to gain the first introduction to a semiconductor device that is based on quantum mechanics. There is a constant interplay between the basic physical laws and the practical development of a commercial device. This philosophy will continue in the understanding of more complicated devices.

PROBLEMS

1. Assume that the cross-sectional areas of a *pn* junction are A_p and A_n. Find the ratio of the ionized donors and acceptors if the depletion widths are x_p and x_n, respectively. Can you say anything about the doping if these two widths are equal?

2. Find the value of the contact potential at $T = 300$ K across a silicon junction if the *p* region has 10^{16} acceptors/cm³ and the *n* region has (a) 10^{14}, (b) 10^{16}, and (c) 10^{18} donors/cm³.

3. Calculate the total depletion width W and the deletion widths in each region of a *pn* junction at room temperature. The dopant densities are $N_a = 10^{15}$/cm³ and $N_d = 10^{17}$/cm³ and $n_i = 10^{10}$/cm³. The relative dielectric constant $\kappa_s = 12$.

4. Repeat Problem 3 if $A_n = 2A_p$ in the depletion region.

5. Find the depletion width x_n in the *n* region if the charge is linearly distributed, the maximum charge density being at the junction. You may assume that the depletion width in the *p* region $x_p \approx 0$ and the contact potential is ϕ_0. Sketch the spatial variation of charge, electric field, and potential in the depletion region. What does this assumption impose on the doping densities?

6. Find the capacitance per unit area of the diode described in Problem 5 and sketch its dependence on applied potential.

7. Show that, if $N_d(x)$ is a slowly varying function of x for the "single-sided" diode described in Problem 5, then it is possible to find this spatial variation from

$$N_d = \frac{c^3}{q\kappa_s}\left(\frac{dC}{d\phi}\right)^{-1}$$

where ϕ is the reverse-bias voltage that is applied across the junction and C is the capacitance per unit area of the unbiased diode.

8. Accurately plot the current (I/I_0) versus voltage (in units of $k_B T/q$) as predicted by the diode equation (19). Also, accurately plot the current due to majority carriers.

9. Show that the distribution of excess holes in the *n*-type semiconductor is approximately linear if $x_n \ll L_h$. Estimate the resulting flux of minority carriers under these conditions.

10. (a) Zener breakdown occurs when the maximum electric field approaches a critical value E_0. If the density on the *p* side is N_a, find the required donor concentration on

the n side to obtain Zener breakdown at a voltage ϕ_B. Use a one-sided step approximation. (b) Evaluate this density if $E_0 = 10^6$ V/cm, $N_a = 10^{20}$/cm^3, and $\phi_B = 2$ V.

11. For a metal–semiconductor diode, what slopes will a curve of $\log(C)$ versus $\log(\phi)$ have, where ϕ is the applied potential, if the charge distribution is (a) constant and (b) proportional to distance?

12. Show that the minority charge distribution for a finite-size diode will involve hyperbolic sines and cosines. Examine this behavior for small-length diodes.

13. The current output from an optical diode is given by $I = I_0 \cos(\pi\phi/2\phi_0)$. Find the maximum power output of the diode.

Chapter Seven

Bipolar Transistors

The revolution in the physics community during the depths of the depression had a profound impact on society in the period of time during and following the worldwide calamity known as World War II. Humanity could no longer claim to be innocent and scientists could no longer hide behind the cloaks of noninvolvement. The mysteries of the atom had been opened and science had undergone a period of euphoria and also serious trepidation as the investigations continued in the ivory tower laboratories. It was as if there was an ongoing period of communion with the inhabitants of the heavens and the secrets were being laid bare in front of their eyes.

The mysteries of the atom had intrigued scientific minds ever since the days of antiquity. This interest was not lessened by decrees or wars, and the period following the development of quantum mechanics certainly did not bring on a .dark age in science and technology. For example, mankind was soaring into the heavens, diseases were being conquered, and the announcement that Bardeen, Brattain, and Shockley had developed a "semiconductor triode" would cause the revolution that had already engulfed physics to spread to the field of electronics. The electrical manufacturing industry would never again be the same! It appeared that even the most common of individuals could start an industry that would soon surpass the resources of the old patrician behemoths, some of which appeared to be headed in the same direction that diplodocus and his fellow dinosaurs had previously trod. To help appreciate this discovery and its subsequent development, the reader will have to first understand the operation of the most fundamental of these new electronic components, the *transistor*. Once this is understood, then the modern-day miracle of taking a room full of older computers and gadgets and putting them on one's wrist will not seem so mysterious.

I. Transistor Operation

In its simplest form, the transistor consists of two *pn* diodes connected to each other such that either the two *n* regions or the two *p* regions are connected together. The two central regions are in actuality the same *n*- or *p*-type semiconductor. For both configurations, one diode is forward biased and the other diode is reverse biased with external voltage sources. The arrangement of

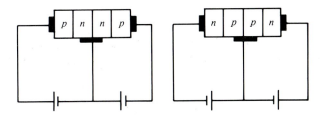

Figure 7.1   Elementary models using *pn* diodes for a *pnnp* and an *nppn* transistor.

the diodes allows us to call the first configuration *pnnp* and the second *nppn*. See Figure 7.1. If now one of the diodes of either configuration is forward biased, a large flow of majority carriers from the outer semiconductor will be injected into the inner region, where they become minority carriers. Let us call the outer region of this diode an *emitter*. If now the central region is small in size, these minority carriers will reach the second diode region, which will be reverse biased. The outer semiconductor will be called the *collector* since it collects these minority carriers that have traversed over the region that is called the *base*.

The unique feature of this configuration is that the voltage that is applied between the emitter and the base controls the current that flows between the base and the collector. Is this model correct? Yes and no. The central semiconductor is not connected with a wire. Rather it is one continuous material and the other semiconductors are implanted in the host as shown in Figure 7.2.

In the bipolar transistor shown in this figure, the actual transistor action can be considered to occur in a "one-dimensional" segment directly underneath the

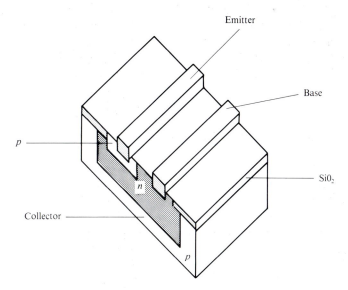

Figure 7.2   Idealized bipolar transistor.

region containing the electrode labeled as the emitter, which is a heavily doped p^+ region. The narrow region, which is an *n*-type semiconductor, is the base, and the lightly doped region beneath it acts as the collector. An insulator covers the transistor and separates the two electrodes on the top. The emitter–base junction is normally forward biased, and the collector–base junction is normally reverse biased.

The transistor shown in Figure 7.2 is a *pnp transistor*. If the *n* and *p* materials were interchanged with *p* and *n* materials, the transistor would have the acronym of an *npn transistor*. The voltage polarities of the biasing batteries would have to be reversed for this device as the emitter–base junction is still to remain forward biased and the base–collector junction remains reverse biased. For this configuration, the emitter region would be a heavily doped n^+ region. We could analyze either of the two models and predict the behavior of the transistor. A summary of the various modes of operation caused by the bias orientations will be discussed in Section III. It is, however, more intuitive to examine the *pnp* transistor since the flow of holes and the current will be in the same direction.

In Figure 7.3(a), an idealized *pnp* transistor in thermal equilibrium is shown. Thermal equilibrium in this case will be defined as connecting all three terminals to a common point, which may be at ground potential. The shaded regions correspond to the widths of the depletion layers at the two *pn* junction regions. The charge density distribution within the transistor is shown in Figure 7.3(b). The three regions of the semiconductor are doped with a relative doping content that satisfies the relation

Emitter doping > base doping > collector doping

In Figure 7.3(c), the energy levels of the transistor are shown. Since the transistor

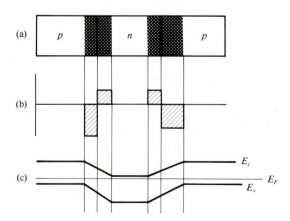

Figure 7.3   Transistor in thermal equilibrium. (a) *pnp* transistor. (b) Charge distribution in the transistor. (c) Transistor energy levels.

is initially assumed to be in equilibrium, the Fermi energy in the three regions lines up. Also, the emitter is more heavily doped than the collector; hence the valence energy level in the emitter will be closer to the Fermi level than the corresponding energy level in the collector.

Let us now examine the transistor under conditions of normal biasing (emitter–base forward biased and collector–base reverse biased). Since the base is common to both the input and output circuits, this configuration is called a *common-base configuration*. This is shown in Figure 7.4 with the same notation as in Figure 7.3.

Certain general features are readily apparent. The depletion widths change due to the bias voltages. The emitter–base width has decreased due to the forward bias and the collector–base width has increased due to the reverse bias. The energy band diagram, which is shown in Figure 7.4(c), exhibits some noticeable changes under conditions of the biasing. Since the emitter–base junction is forward biased, holes are readily injected from the heavily doped emitter into the base region. Correspondingly, electrons are injected from the base region into the emitter. In an ideal diode, the flow of these two particles constitutes the emitter current. The bias across the collector–base junction is reverse biased, and only a small reverse saturation current will flow across this junction. Recall from the previous chapter that this current could be in the range of microamperes, while the forward-biased current would be in the range of milliamperes. In practice, the base region will be made small such that the injected holes from the emitter can easily diffuse across the base region and reach the base–collector junction. These holes will then act like bubbles and "float up" to the collector. If the base region is sufficiently small, the current due to these diffused holes could be much larger than the small reverse saturation current that already exists in the reverse-biased base–collector junction. This process of controlling the current flow in the base–collector region with a changing emitter–base voltage is called the *transistor action*.

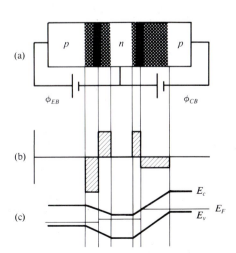

Figure 7.4   A *pnp* transistor under conditions of normal biasing. (a) *pnp* transistor. (b) Charge distribution in the transistor. (c) Transistor energy levels.

In Figure 7.5, the currents that exist in a *pnp* transistor are shown. Kirchhoff's laws apply and we can consider the transistor as a node. The current that enters the transistor at the emitter terminal (E) divides itself into the current that leaves the collector terminal (C) and the current that leaves the base terminal (B).

$$I_E = I_B + I_C \tag{1}$$

The other currents are internal to the transistor. The current I_C from the collector terminal is comprised of the current that will make it to the collector–base junction $I_{CB(j)}$, the electron current due to the electrons migrating from the collector to the base I_{Cn}, and the fraction of the emitter current that has made it across the base region from the emitter. This last component will be written as

$$I_{BC(j)} = \alpha_F I_{EB(j)} \tag{2}$$

where α_F is called the *base transport factor*. The current I_C is therefore

$$I_C = -I_{CB(j)} + I_{BC(j)} + I_{Cn} = -I_{CB(j)} + \alpha_F I_{EB(j)} + I_{Cn} \tag{3}$$

The *common base current gain* α_0 is defined as

$$\alpha_0 = \frac{I_{Cp}}{I_{Ep}} \tag{4}$$

where only the current due to the migration of holes in the collector current for a *pnp* transistor is included in this definition. The emitter and collector currents include both the electron and the hole contributions:

$$I_E = I_{En} + I_{Ep}, \qquad I_C = I_{Cp} + I_{Cn}$$

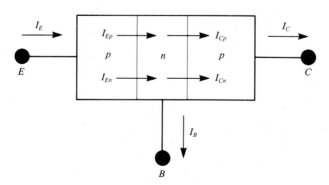

Figure 7.5   Various currents in a *pnp* transistor.

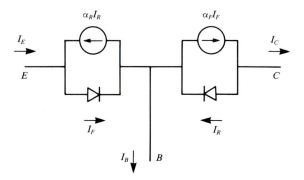

Figure 7.6   Equivalent circuit model of a transistor.

The current I_E from the emitter terminal is comprised of the current that will make it to the base–emitter junction $I_{BE(j)}$, the electron current I_{En} created by the electrons traversing from the base to the emitter, and that fraction of the current that has made it across the base region from the collector $\alpha_R I_{CB(j)}$, where α_R is a constant called the *reverse base transport factor*. The current I_E is given by

$$I_E = I_{EB(j)} + I_{En} - \alpha_R I_{CB(j)} \tag{5}$$

It is now possible to suggest an equivalent circuit model for the transistor defined by Equations (3) and (5). This is shown in Figure 7.6. In this figure, the diodes are assumed to be ideal diodes that have reverse saturation currents of I_{0E} and I_{0C}, respectively, for the emitter and the collector. The important feature of this model that should be emphasized is the *voltage-dependent current generators* that appear in each side of the base region. This circuit model is called an Ebers–Moll model for the transistor. Other circuit models will be described later.

II. Charge Behavior in the Base Region

The base region of a transistor is in juxtaposition with both the emitter and the collector. Charge that emanates from the emitter passes through this region on its path to the collector. During this period of motion, it is a minority carrier, p_n. As the transistor depends on its behavior there, it is incumbent for the serious student to examine the behavior of the minority charge in the base region in some detail. This will be investigated in this section.

It is advantageous to initially make certain simplifying assumptions that maintain the integrity of the calculation but still illustrate the operation of the device. These assumptions are the following:

1. The doping of the device is uniform in each region.
2. There is a low level of injection.
3. There is no generation or recombination in the depletion regions.

4. Each region has infinite conductivity, so the series resistance within the semiconductor can be neglected.

These assumptions make the problem linear, and the resulting calculation will be very similar to finding the charge distribution in a *pn* diode.

In the steady state, the perturbation of the minority carrier distribution Δp_n is governed by the equation

$$D_p \frac{d^2 \Delta p_n}{dx^2} - \frac{\Delta p_n}{\tau_p} = 0 \qquad (6)$$

where D_p and τ_p are the diffusion coefficient and lifetime for the holes in the base region, respectively. The perturbation of the minority hole density Δp_n is given by

$$\Delta p_n = p_n - p_{n0}$$

where p_{n0} is a small equilibrium value of minority carriers in the base region and p_n corresponds to the holes that are injected into the base. The solution of (6) is

$$\Delta p_n = A e^{x/L_p} + B e^{-x/L_p} \qquad (7)$$

where $L_p = [D_p \tau_p]^{1/2}$. This is similar to the solution obtained for the case of *pn* diodes. The constants of integration, A and B, are determined from the boundary conditions at $x = 0$ and at $x = W$, where W is the width of the base region. The location $x = 0$ will be taken to be at the emitter–base junction where the perturbation in the minority carrier density is expressed as

$$\Delta p_{n0} = p_{n0} [e^{q\phi_{EB}/(k_B T)} - 1] \qquad (8)$$

This states that under forward bias the minority carrier concentration is exponentially increased at the edge of the emitter–base junction. At the collector–base junction, $x = W$, the minority carrier density distribution p_n, which has been injected from the emitter, is zero at the reverse-biased base–collector diode junction. Therefore,

$$\Delta p_n(x = W) = -p_{n0} \qquad (9)$$

Recall that $p_{n0} = n_i^2/N_B$, where N_B is the uniform donor concentration in the base region. With these boundary conditions, (7) becomes

$$p_n(x) = p_{n0} [e^{q\phi_{EB}/(k_B T)} - 1] \left\{ \frac{\sinh[(W - x)/L_p]}{\sinh(W/L_p)} \right\} + p_{n0} \left[1 - \frac{\sinh(x/L_p)}{\sinh(W/L_p)} \right] \qquad (10)$$

If the base region is small such that $x/L_p \le W/L_p \ll 1$, then the hyperbolic functions can be replaced by their arguments ($\sinh \beta \approx \beta$) and (10) becomes

$$p_n(x) \approx p_{n0}e^{q\phi_{EB}/(k_BT)}\left[1 - \frac{x}{W}\right] \tag{11}$$

where the approximation that at room temperature the exponential term is much larger than 1 has been employed.

The minority carrier distribution in a typical *pnp* transistor operated under the normal bias conditions and with the approximation of a narrow base region is shown in Figure 7.7. The procedure for calculating the minority carrier distribution in the emitter and collector regions is similar to the procedure used to calculate the distribution in the base region. The boundary conditions that should be imposed at the emitter–base interface are

$$n_E = n_{E0}e^{q\phi_{EB}/(k_BT)}$$

and at the collector–base interface

$$n_C = n_{C0}e^{-q|\phi_{CB}|/(k_BT)} = 0$$

where n_{E0} and n_{C0} are the equilibrium electron densities in the emitter and the collector, respectively. These two expressions are valid descriptions at the edge of the depletion layers at $x = -x_E$ and at $x = x_C$, which extend into the emitter and the collector regions, respectively. The boundary conditions at the far edges of the emitter and the collector regions are such that the perturbation in the density approaches zero, which is valid in the limit that the diffusion lengths L_E and L_C are much smaller than the widths of the emitter and the collector, respectively. If we impose these boundary conditions for the solution of (6), the solution for the densities in these two regions is obtained:

$$x \le -x_E: \quad n_E(x) = n_{E0} + n_{E0}[e^{q\phi_{EB}/(k_BT)} - 1]e^{(x+x_E)/L_E} \tag{12}$$

$$x \ge x_C: \quad n_C(x) = n_{C0} - n_{C0}e^{-(x-x_C)/L_C} \tag{13}$$

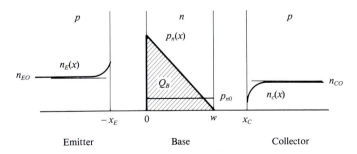

Figure 7.7 Minority carrier distribution in a *pnp* transistor.

The total excess minority charge that is found in the base region Q_B can be computed from the integral

$$Q_B = qA \int_0^W \Delta p_n(x)\, dx \qquad (14)$$

where A is the cross-sectional area of the transistor. This can be easily integrated if we assume that $p_n(x = 0) \gg p_{n0} \approx 0$ (i.e., a triangular approximation for the density as shown in Figure 7.7). In this case, we obtain

$$Q_B \approx \frac{qAW p_n(x = 0)}{2} \qquad (15)$$

Now that the minority carrier distributions are known, it is possible to calculate the currents since they are related to the density distribution through the derivative operation and the appropriate diffusion coefficient. For example, the hole current I_{Ep} of holes injected into the base region at $x = 0$ is found from

$$I_{Ep} = -AqD_p \left.\frac{dp_n}{dx}\right|_{x=0} \qquad (16)$$

Note the appearance of a minus sign in this equation. The flow of holes will be in the direction of decreasing minority hole density or in this case in the direction of increasing position x. The distribution given in (10) is substituted into this equation, and we compute

$$I_{Ep} = \frac{qAD_p p_{n0}}{L_p} \coth\left(\frac{W}{L_p}\right)\left[e^{q\phi_{EB}/(k_B T)} - 1 + \frac{1}{\cosh(W/L_p)}\right] \qquad (17)$$

In the limit of $W/L_p \ll 1$, (16) can be simplified since the following approximations can be employed for the hyperbolic functions:

$$\coth \beta = \frac{1}{\tanh \beta} \approx \frac{1}{\beta}$$

$$\sinh \beta \approx \beta$$

$$\operatorname{sech} \beta = \frac{1}{\cosh \beta} \approx 1 - \frac{\beta^2}{2}$$

If these are included in (16), we obtain

$$I_{Ep} \approx \frac{qAD_p p_{n0}}{L_p}\left(\frac{L_p}{W}\right)\left[e^{q\phi_{EB}/(k_B T)} - 1 + \left\{1 - \left(\frac{W}{L_p}\right)^2\right\}\right]$$

or

$$I_{Ep} \approx \frac{qAD_pp_{n0}}{W}e^{q\phi_{EB}/(k_BT)} \tag{18}$$

where it has been assumed that the exponential term dominates over all other terms. This agrees with the hole current that we would obtain if (11) had been directly employed in (16).

In a similar fashion, the hole current collected by the collector is computed from

$$I_{Cp} = -AqD_p\frac{dp_n}{dx}\bigg|_{x=W} = \frac{qAD_pp_{n0}}{L_p}\left[\frac{1}{\sinh(W/L_p)}\right]\left[e^{q\phi_{EB}/(k_BT)} - 1 + \cosh\left(\frac{W}{L_p}\right)\right] \tag{19}$$

This can be similarly approximated as

$$I_{Cp} \approx \frac{qAD_pp_{n0}}{W}e^{q\phi_{EB}/(k_BT)} \tag{20}$$

The electron currents I_{En}, which is caused by the electron flow from the base to the emitter, and I_{Cn}, which is caused by the electron flow from the collector to the base, are given by

$$I_{En} = -AqD_E\frac{dn_E}{dx}\bigg|_{x=x_E} = \frac{AqD_En_{E0}}{L_E}[e^{q\phi_{EB}/(k_BT)} - 1] \tag{21}$$

and

$$I_{Cn} = -AqD_C\frac{dn_C}{dx}\bigg|_{x=x_C} = \frac{qAD_Cn_{C0}}{L_C} \tag{22}$$

In these two equations, D_E and D_C are the diffusion coefficients of the emitter and the collector, respectively.

The terminal current I_E is found from

$$I_E = I_{Ep} + I_{En}$$

The collector current I_C is given by

$$I_C = I_{Cp} + I_{Cn}$$

In the limit of $W/L_p \ll 1$, these currents can be written as

$$I_E \approx \left[\frac{qAD_pp_{n0}}{W} + \frac{qAD_En_{E0}}{L_E}\right][e^{q\phi_{EB}/(k_BT)} - 1] + \frac{qAD_pp_{n0}}{W}$$

and

$$I_c \approx \frac{qAD_pp_{n0}}{W}\left[e^{q\phi_{EB}/(k_BT)} - 1\right] + \left[\frac{qAD_pp_{n0}}{W} + \frac{qAD_cn_{C0}}{L_C}\right]$$

An estimate for the collector current can be obtained from (15) with

$$I_c \approx \frac{qAD_p}{W}p_{n0}e^{q\phi_{EB}/(k_BT)} \approx \frac{qAD_p}{W}\frac{2Q_B}{qAW} = \frac{2D_p}{W^2}Q_B \qquad (23)$$

where the small electron current has been neglected. Therefore, with these approximations, we have reached the important conclusion that the collector current is proportional to the minority carrier charge stored in the base.

From this detailed derivation, we note that all the currents that reach the three terminals of the transistor are related to the minority carrier distribution in the base region. The emitter and collector currents are given by the minority carrier gradients at the junction boundaries $x = 0$ and $x = W$. Both of these currents are proportional to the stored charge in the base. The applied voltages control the densities at these boundaries through the term $e^{q\phi_{EB}/(k_BT)}$.

Example 7.1   Find the base current of holes in the limit of a very thin base region; that is, $W/L_p \ll 1$.

Answer: The minority carrier distribution in this limit is given by (11), which is

$$p_n(x) \approx p_{n0}e^{q\phi/(k_BT)}\left[1 - \frac{x}{W}\right]$$

The hole current injected from the emitter into the base is given by

$$I_{Ep} = -qAD_p\frac{dp_n}{dx}\bigg|_{x=0} \approx \frac{qAD_pp_{n0}e^{q\phi/(k_BT)}}{W}$$

The hole current absorbed by the collector from the base is given by

$$I_{Cp} = -qAD_p\frac{dp_n}{dx}\bigg|_{x=W} \approx \frac{qAD_pp_{n0}e^{q\phi/(k_BT)}}{W}$$

The base current of holes is computed from Kirchhoff's law as

$$I_{Ep} = I_{Bp} + I_{Cp}$$

or $I_{Bp} \approx 0$. This states that the hole current passes from the emitter directly to the collector and none leaves by the base terminal (i.e., $\alpha_F = 1$). Note also that the ratio of the collector current to the emitter current is independent of the bias voltage applied between the collector and the base.

The *emitter injection efficiency* γ is defined as the ratio of the hole current to the sum of the hole and electron currents, from which we compute

$$\beta = \frac{I_{Ep}}{I_{Ep} + I_{En}} \approx \frac{D_p p_{n0}/W}{\dfrac{D_p p_{n0}}{W} + \dfrac{D_E n_{E0}}{L_E}} = \frac{1}{1 + \dfrac{D_E}{D_p}\dfrac{n_{E0}}{p_{n0}}\dfrac{W}{L_E}} = \frac{1}{1 + \dfrac{D_E}{D_p}\dfrac{N_B}{N_E}\dfrac{W}{L_E}} \tag{24}$$

where the impurity doping in the base is $N_B = n_i^2/p_{n0}$ and the impurity doping in the emitter is $N_E = n_i^2/n_{E0}$. To improve the efficiency γ, we should use an emitter that is more heavily doped than the base ($N_E \gg N_B$).

There is one remaining design consideration for the physical operation of a bipolar transistor; it is the maximization of the *base transport factor* α_F, which is defined as

$$\alpha_F = \frac{I_{Cp}}{I_{Ep}} = \frac{\dfrac{qAD_p p_{n0}}{L_p}\dfrac{1}{\sinh(W/L_p)}[e^{q\phi_{EB}/(k_B T)} - 1] + \cosh(W/L_p)}{\dfrac{qAD_p p_{n0}}{L_p}\coth\left(\dfrac{W}{L_p}\right)[e^{q\phi_{EB}/(k_B T)} - 1] + \dfrac{1}{\cosh(W/L_p)}}$$

$$\approx \frac{1}{\cosh(W/L_p)} \approx 1 - \frac{W^2}{2L_p^2} \tag{25}$$

where the assumption that the exponential term dominates all the other terms has been made. This factor is maximized if the ratio W/L_p is minimized. In practice, this means minimizing the width of the base region W, since L_p is almost constant. This result could probably have been predicted from an inspection of the transistor itself. The current I_B that could escape through the base will be minimized if the area of the base is minimized; this states that the width of the base W should be as small as possible. Hence there is almost a direct path from the emitter to the collector for the flow of the holes.

III. Transistor Modes of Operation

There are four modes of operation for the bipolar transistor. These modes depend on the polarity of the bias voltages that are applied between the emitter–base and the collector–base junctions. These biases apply to both the *pnp* and the *npn* transistor configurations. The modes are summarized in Table 7.1.

The active mode is the normal mode of operation and its operation was described in the previous section. In the saturation mode, both junctions are forward biased, which reduces both depletion widths. This creates a transistor that is in a conducting state and allows the transmittance of large currents (i.e., a closed switch). The cutoff mode has both junctions reverse biased, which increases both depletion widths. The charge stored in the base region is small, and

TABLE 17.1

Mode	Emitter–Base Bias	Collector–Base Bias
Active	Forward	Reverse
Saturation	Forward	Forward
Cutoff	Reverse	Reverse
Inverted	Reverse	Forward

this reduces the collector current (i.e., an open switch). The inverted mode acts as a "poor" active transistor due to the low doping of the collector with respect to the base. The current gain will be small in this mode of operation.

In operating the bipolar transistor in the cutoff mode, one restriction must be imposed on the minimum width W of the base region. The width must be sufficiently wide that the depletion widths at the two junctions do not overlap. This overlapping is sometimes called "punch through" and can lead to deleterious results.

IV. Transistor Characteristics

In Section II, the currents that exist in a bipolar transistor that was operated in the common-base configuration were derived in terms of the charge that was stored in the base region. This derivation is valid for moderate voltages but, as is noted in experiments, there are some significant deviations for voltages that are either very large or very small.

A typical measured output voltage–current characteristic is shown in Figure 7.8 for a *pnp* transistor that is operating in the common-base configuration. Certain general features of the transistor operation become apparent from these characteristics. First, the various modes of normal operation that were described in Section III are indicated in the figure. The inverted mode is normally not employed. In the active mode of operation, the ratio of the collector current to the emitter current is approximately 1, and it is almost independent of the bias potential ϕ_{BC} over a significant range of voltages. This is in agreement

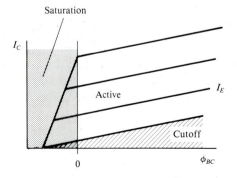

Figure 7.8 Transistor characteristics of a *pnp* transistor. The higher values of collector current occur for higher values of emitter current.

with the ratio of equations (18) and (20) for the hole currents. As the electron currents have a similar dependence on voltage, the ratio of the total currents will yield similar conclusions. As the value of ϕ_{BC} approaches zero, holes are still extracted from the base region by the collector. To reduce this flow, a small forward bias must be applied to the base–collector junction such that it is in the saturation mode. This forward bias will increase the hole density at $x = W$ to make it equal to the hole density at the emitter–base junction at $x = 0$. This reduces the value of $d(\Delta p_n)/dx$ at $x = W$, which in turn reduces the hole current there.

In this chapter, we have considered only the operation of the transistor in the common-base configuration. A more frequently used configuration is the *common-emitter* configuration in which the emitter is common to the input and the output circuits. Such a configuration is shown in Figure 7.9. The output transistor characteristics for a *pnp* transistor operating in this mode are also shown in this figure.

It is possible to relate the common-emitter and the common-base configurations as shown later. Prior to effecting this relationship, let us return to the common-base model and define a few more terms. The *common-base current gain* α_0 is defined as $\alpha_0 = I_{Cp}/I_E$, where

$$\alpha_0 = \frac{I_{Cp}}{I_{Ep} + I_{En}} = \frac{I_{Ep}}{I_{Ep} + I_{En}}\left(\frac{I_{Cp}}{I_{Ep}}\right) \tag{26}$$

The first term is called the *emitter injection efficiency* γ and the second term is the *base transport factor* α_F. This is the same as (2) with the added assumption that all the holes that actually make it to the junction actually appear as holes at the terminal. This simplifies the model somewhat and is not too bad an approximation.

From this, one concludes that the coefficients are related by

$$\alpha_0 = \gamma\alpha_F \tag{27}$$

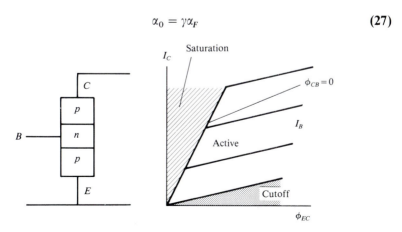

Figure 7.9   Common-emitter transistor characteristics.

For a well-designed transistor, all three of these coefficients are close to unity in value. The component of electron current I_{En} is usually small in comparison with the hole current I_{Ep} for a *pnp* transistor.

The collector current can be expressed in terms of the emitter current I_E via

$$I_C = I_{Cp} + I_{Cn} = \alpha_F I_{Ep} + I_{Cn} = \gamma \alpha_F \frac{I_{Ep}}{\gamma} + I_{Cn} = \alpha_0 I_E + I_{Cn} \qquad (28)$$

It is common to designate I_{Cn} as I_{CBO} to reflect the fact that this current flows between the collector and the base. The third subscript describes the state of the third terminal with respect to the second. This current is the leakage current between the collector and the base with the emitter left open. Therefore,

$$I_C = \alpha_0 I_E + I_{CBO} \qquad (29)$$

If we now substitute (1) in (29), the following results:

$$I_C = \alpha_0 (I_B + I_C) + I_{CBO}$$

Solving for I_C, we obtain

$$I_C(1 - \alpha_0) = \alpha_0 I_B + I_{CBO} \qquad (30)$$

This can be written as

$$I_C = \frac{\alpha_0}{1 - \alpha_0} I_B + \frac{I_{CBO}}{1 - \alpha_0} \qquad (31)$$

The first term in (31) is called the *common-emitter current gain*. It is typically given the symbol β_0, where

$$\beta_0 = \frac{\Delta I_C}{\Delta I_B} = \frac{\alpha_0}{1 - \alpha_0}$$

As α_0 is close to unity, the value of β_0 may be large. With the emitter injection efficiency γ being close to unity, we can write

$$\beta_0 = \frac{\alpha_0}{1 - \alpha_0} = \frac{\gamma \alpha_F}{1 - \gamma \alpha_F} \approx \frac{\alpha_F}{1 - \alpha_F} = \frac{2L_p^2}{W^2} \qquad (32)$$

where (25) and the ratio $W/L_p \ll 1$ have been employed.

Example 7.2   Show that a *pnp* transistor can be used for amplification in the common-emitter circuit.

Answer: The dc operating characteristics are given by the currents in the two circuits. The current from the base is $I_B = 6$ V/60 k$\Omega = 10^{-4}$ A. The current from the collector is $I_C = 12$ V/600 $\Omega = 0.02$ A. Hence the ratio of the two dc currents is 200. If in addition to this dc current a small ac signal is introduced in the base circuit, the output current will be amplified by this ratio of 200.

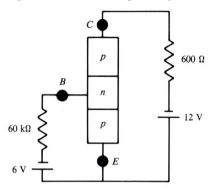

Further detailed treatment of various circuit configurations using a transistor is better left to a text dedicated to electronic circuits. However, the basic hybrid circuit that is often used to model a transistor is described in Section VI.

V. Switching Transients

Digital electronics has imposed new requirements for the operation of electronic circuits in that the transistor must be able to be switched from one state of operation to the other in a very short period of time. For a bipolar junction transistor, this will involve the passing from the cutoff mode, through the active mode, and into the saturation mode, or in the reverse sequence of transition. This change cannot be done instantly but requires a finite time, as the charge in the base region has to be either "swept out" or charge has to be introduced into this region. If we examine the minority charge distribution that was calculated and displayed in Figure 7.7, we realize that the switching times will be dictated by the characteristic times of the minority charge motion in the base region. This is particularly true if the emitter and collector are heavily doped. The minority carrier concentration in the base region will then be larger than in the other two regions.

In Figure 7.10, the consequence of injecting a current pulse into the base region of a transistor that was previously "turned off" is shown. The current pulse is injected at a time $t = 0$, and this will constitute the "turning on" of the transistor and its properties will be examined. The total excess minority carrier charge stored in the base region at any particular time is given by

$$Q_B = qA \int_0^W \Delta p_n(x)\, dx \qquad (33)$$

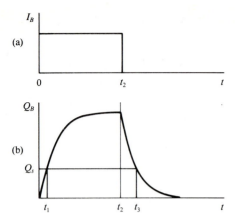

Figure 7.10 Switching
characteristics of a transistor.
(a) Applied current pulse to the base.
(b) Charge stored in the base.

The time variation of the charge density is governed by the continuity equation

$$-\frac{1}{q}\frac{\partial j_p}{\partial x} = \frac{\partial \Delta p_n}{\partial t} + \frac{\Delta p_n}{\tau_p} \qquad (34)$$

where τ_p is the excess minority carrier lifetime and j_p is the hole current density.
Integrating (34) from $x = 0$ to $x = W$ and using (33), we obtain

$$I_p(0) - I_p(W) = \frac{dQ_B}{dt} + \frac{Q_B}{\tau_p} \qquad (35)$$

The difference of the two terms on the left side of this equation is just the current
that leaves the base terminal i_B. Hence (35) can be written as

$$i_B = \frac{dQ_B}{dt} + \frac{Q_B}{\tau_p} \qquad (36)$$

At this point, we could have employed the common circuit model, which states
that all currents "enter" the transistor, rather than the intuitive approach used
here.

After the transistor is switched ($t > 0$), the base current is constant with a
value I_B. The solution of (36) is therefore given by

$$Q_B(t) = I_B\tau_p(1 - e^{-t/\tau_p}) \qquad (37)$$

where the assumption is used that there is no stored charge in the base region
prior to switching; that is, $Q_B(t < 0) = 0$. For long times, which are defined as
$t \gg \tau_p$, this charge Q_B approaches $I_B\tau_p$. This value may be greater than the
saturated value Q_S of the transistor.

The collector current can be calculated starting from (23) as

$$I_C \approx \frac{2D_p}{W^2} Q_B(t) \approx \frac{2D_p}{W^2} I_B \tau_p (1 - e^{-t/\tau_p}) \tag{38}$$

This can be simplified further if we can estimate the time τ_B that it takes for a hole to traverse across the base region. This can be found from the following integral of the width of the base region divided by the local space-dependent hole velocity:

$$\tau_B = \int_0^W \left(\frac{1}{v(x)} \right) dx \tag{39}$$

where $v(x)$ is the effective hole minority carrier velocity in the base region of the *pnp* transistor. This is related to the hole current and the local hole density through the definition

$$I_p = qv(x)p(x)A$$

Therefore, (39) can be written as

$$\tau_B = \int_0^W \left(\frac{qp(x)A}{I_p} \right) dx$$

Using the straight-line approximation for the charge distribution, this can be integrated using (11) and (20) as

$$\tau_B = \int_0^W \frac{qAp_{n0}e^{q\phi_{EB}/(k_BT)}[1 - (x/W)]}{(qAD_p/W)p_{n0}e^{q\phi_{EB}/(k_BT)}} dx$$

$$= \int_0^W \frac{W}{D_p} \left[1 - \frac{x}{W} \right] dx$$

The transit time τ_B is obtained finally as

$$\tau_B = \frac{W^2}{2D_p} \tag{40}$$

This result could have been postulated from dimensional arguments also. The diffusion constant D_p has the dimensions of (length)^2/(time). A characteristic length of the base region is W, and a characteristic time for charge motion in the base region is τ_B. Hence $D_p = W^2/\tau_B$. The factor of 2 can be included since an average value is required.

The collector current can therefore be written as

$$I_C = I_B \frac{\tau_p}{\tau_B}(1 - e^{-t/\tau_p}) \tag{41}$$

For values of $Q_B > Q_S$, the device is operated in the saturation mode, and both the emitter and collector currents remain essentially constant. Let us call this time $t = t_1$ when the charge in the base region is just equal to Q_S. Therefore, from (37)

$$Q_S = I_B \tau_p (1 - e^{-t_1/\tau_p}) \tag{42}$$

and this time can be found from

$$t_1 = \tau_p \ln\left[\frac{1}{1 - (Q_S/I_B\tau_p)}\right] \tag{43}$$

The values of Q_S and I_B can be estimated from external circuit elements as shown in Figure 7.11, which is a circuit diagram for a transistor switching circuit. We estimate these values as

$$Q_S \approx \frac{\phi_{cc}}{R_L} \tau_B \tag{44}$$

and

$$I_B \approx \frac{\phi_s}{R_s} \tag{45}$$

This "turn-on" switching time t_1 can be reduced if the minority lifetime τ_p or saturation value of charge Q_S is made smaller or the value of the base current I_B is increased.

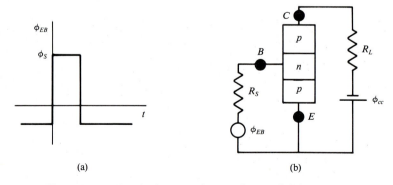

(a)   (b)

Figure 7.11   Circuit diagram of a transistor switching circuit.

The "turn-off" transient is calculated in a similar manner. We will assume that the base current is switched to zero at a time $t = t_2$. To calculate the charge behavior, we again have to solve (36) with the base current $i_B = 0$. The solution of this is

$$Q_B(t) = Q_B(t_2)e^{-(t - t_2)/\tau_p} \tag{46}$$

which is valid for $t \geq t_2$. Again, as shown in Figure 7.10, the charge stored in the base Q_B will equal the saturated value Q_S at a time $t = t_3$. During the time interval $t_3 - t_2$, the transistor is in the saturation mode and the collector current stays constant. This time interval is called the *storage time delay* t_s. The collector current during this time interval is given by

$$i_c = \frac{Q_B(t - t_2)}{\tau_B} = \frac{Q_B(t_2)}{\tau_B}e^{-(t - t_2)/\tau_p} \tag{47}$$

The storage time t_s is given by

$$t_s = t_3 - t_2 = \tau_p \ln\left[\frac{Q_B(t_2)}{Q_S}\right] \tag{48}$$

If t_2 is much larger than τ_p, Q_B is approximately equal to $I_B \tau_p$ from (37). Therefore,

$$t_s \approx \tau_p \ln\left[\frac{I_B \tau_p}{Q_S}\right] \tag{49}$$

Again, this time will be reduced if τ_p is reduced.

Once again we have used the continuity equation for the minority carrier charge distribution in the base region in order to derive the transient switching characteristics for the bipolar transistor. The turn-on time depends on how fast we can add holes to the base region of the *pnp* transistor where they become minority carriers. The turn-off time depends on how fast these holes can be removed by recombination. For faster switching characteristics, the lifetime τ_p should be reduced. One method of doing this is to introduce efficient generation–recombination centers near the midgap of the energy levels of the base region.

In this derivation of the transient characteristics for the *pnp* transistor, we calculated the time τ_B that it takes for the hole to traverse the base region (39). This time imposes a high-frequency limitation for the response of the bipolar transistor.

VI. Transistor Equivalent Circuits

The bipolar transistor has been extensively used in a variety of circuit configurations and various models have been brought forth to describe its operation. One such moodel is the Ebers–Moll model, which was described previously. There are other circuit models that are based on the small-signal operation of the transistor. One of these is the *hybrid* model, which will be described in this section. By starting with the assumption of small-signal operation, the reader will obtain a model that assumes that all elements are linear, a model that has received considerable attention in electronics.

Consider the common-base configuration as shown in Figure 7.12. The transistor is treated as a two-port element where two of the four transistor variables i_E, i_C, ϕ_{EB}, and ϕ_{CB} may be treated as independent variables and the other two as dependent variables that are functions of these independent variables. The hybrid model treats the input current i_E and the output voltage ϕ_{CB} as the independent variables, and we can write

$$\phi_{EB} = f_1(i_E, \phi_{CB}) \tag{50}$$

$$i_C = f_2(i_E, \phi_{CB}) \tag{51}$$

where f_1 and f_2 indicate the functional dependence. It is possible to manipulate the equations that were derived in Section II to obtain the functional dependencies directly. However, since a linear model is sought to represent the transistor, it is easier to examine the transistor characteristics directly.

The functional forms written in (50) and (51) can be expanded in a Taylor series as

$$\phi_{EB} = \frac{\partial \phi_{EB}}{\partial i_E} i_E + \frac{\partial \phi_{EB}}{\partial \phi_{CB}} \phi_{CB} \tag{52}$$

$$i_C = \frac{\partial i_C}{\partial i_E} i_E + \frac{\partial i_C}{\partial \phi_{CB}} \phi_{CB} \tag{53}$$

In writing (52) and (53), we have assumed that only the small-signal variation about a fixed bias point (or a "quiescent operating point"; given the acronym "Q point") is of interest in this derivation. Also, the use of partial derivatives implies that the other dependent variable is held constant during the differentiation. Note also that two of the differentiated terms are dimensionless and two have the dimensions of an impedance or an admittance.

The derivatives can be evaluated graphically from the transistor characteristics indicated in Figure 7.8, and we can write (52) and (53) as a set of linear

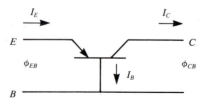

Figure 7.12 Common-base configuration of a *pnp* transistor.

equations:

$$\phi_{EB} = h_{ib}i_E + h_{rb}\phi_{CB} \tag{54}$$

$$i_C = h_{fb}i_E + h_{ob}\phi_{BC} \tag{55}$$

Each term of this model carries a name, and we call the procedure "finding the h parameters" for a common-base configuration of a transistor. The h parameters can be identified by comparing the terms in (52) and (53) with those in (54) and (55), and they are given the following names.

h_{ib} = common-base input impedance

h_{rb} = reverse voltage transverse function

h_{fb} = common-base forward current transfer ratio

h_{ob} = output admittance

With the linear representation for the transistor in the common-base configuration given in (54) and (55), it is possible to model the transistor with the equivalent circuit shown in Figure 7.13. This model is valid at low frequencies. As the frequency of operation is increased, we would have to incorporate transit time effects and diffusion capacitances. This can be modeled with additional capacitors in parallel with the resistors. This, of course, increases the complexity of the model.

The same procedure as just outlined can be employed to determine an equivalent circuit for a *pnp* transistor operating in the common-emitter configuration shown in Figure 7.14. In this configuration, the dependent variables are the input voltage ϕ_{BE} and the collector current i_C. The input current i_B and the output voltage ϕ_{CE} are the dependent variables. Therefore,

$$\phi_{BE} = f_3(i_B, \phi_{CE}) \tag{56}$$

$$i_C = f_4(i_B, \phi_{CE}) \tag{57}$$

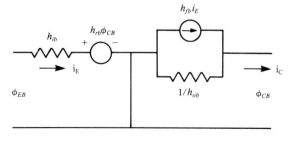

Figure 7.13   Common-base h-parameter representation of a *pnp* transistor.

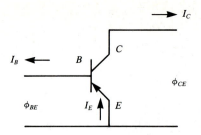

Figure 7.14   Configuration of a common-emitter *pnp* transistor circuit.

where f_3 and f_4 are functional dependencies. These two equations for ϕ_{BE} and i_C can be expanded about the quiescent operating point, and we can write

$$\phi_{BE} = h_{ie}i_B + h_{re}\phi_{CE} \tag{58}$$

$$i_C = h_{fe}i_B + h_{oe}\phi_{CE} \tag{59}$$

where

$$h_{ie} = \frac{\partial \phi_{BE}}{\partial i_B} = \text{common-emitter input impedance}$$

$$h_{re} = \frac{\partial \phi_{BE}}{\partial \phi_{CE}} = \text{common-emitter reverse voltage transfer ratio}$$

$$h_{fe} = \frac{\partial i_C}{\partial i_B} = \text{common-emitter forward current transfer ratio}$$

$$h_{oe} = \frac{\partial i_C}{\partial \phi_{CE}} = \text{output conductance}$$

The values for each of these partial derivatives can be found from Figure 7.9 by graphically differentiating the curves at a quiescent operating point (*Q* point). An equivalent circuit is shown in Figure 7.15.

Other models for the transistor exist, and they can be found in a similar fashion by taking different combinations of voltages and currents as the

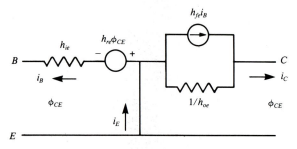

Figure 7.15   Equivalent circuit of a *pnp* transistor in the common-emitter configuration.

dependent and independent variables. This topic is better left for an electronic circuits course, where it is dealt with in more detail and more applications are vigorously presented.

VII. Conclusion

The fundamentals of transistor operation can be determined from a consideration of the characteristics of the minority carrier charge distribution in the base region. Once this distribution is known, the currents can be calculated, since they just diffuse to the less dense region. In the common-emitter mode of operation, small changes in the base current can cause large changes in the collector current, which implies that it can be used as an amplifier. The external bias supplies determine the quiescent operating point of the transistor. Signal generators can switch the transistor operation from a conducting mode of operation to a nonconducting mode, the switching time being governed by the lifetime of the minority carriers in the base region. They can also provide a signal that is to be amplified by the transistor amplifier.

PROBLEMS

1. Show that Equation (11) follows from (10) by carrying forth the detailed derivation.
2. Plot equation (10) to ascertain the charge distribution for different values of $W/L_p = 0.1$, 1, and 10.
3. Evaluate the charge stored in the base region for $W/L_p = 0.1$, 1, and 10.
4. Derive the expressions for the current in the limit of $W/L_n \ll 1$ at the various terminals for an npn transistor.
5. Find the emitter efficiency for a pnp transistor in the limit of $W/L_p \ll 1$.
6. For an npn transistor, draw circuit diagrams including the bias voltages for four modes of transistor operation. Clearly indicate the behavior of the depletion widths at each junction as compared with the unbiased mode.
7. Let us approximate the transistor characteristics for a common-base configuration with the following expression:

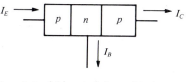

$$I_C \approx I_E \tanh(\phi_{BC} + \phi_0) \qquad \text{for } \phi_{BC} \geq -\phi_0$$

   (a) Sketch the resulting transistor characteristics to ascertain the validity of the approximation.
   (b) Find an expression for the base current.
   (c) Find the voltage ϕ_{BC}, where the base current is $0.1I_E$.

8. Find an equivalent circuit for the *pnp* transistor whose characteristics are approximated with the expression

$$I_C \approx I_E \tanh(\phi_{BC} + \phi_0) \qquad \text{for } \phi_{BC} \geq -\phi_0$$

9. Determine and sketch the transistor characteristics for a common-emitter configuration if the common-base configuration can be approximated with

$$I_C \approx I_E \tanh(\phi_{BC} + \phi_0) \qquad \text{for } \phi_{BC} \geq -\phi_0$$

10. Let the charge stored in the base region of a *pnp* transistor for $t \leq 0$ be Q_{B1}. Derive the expression for the charge stored in the base region if a current I_B is extracted from the base at $t > 0$.

11. In Equation (40), an estimate of the time required for a particle to traverse the base region in the limit $W/L_p \ll 1$ was obtained. Using the next level of approximation for the hyperbolic functions, ascertain if this transit time will increase or decrease if W/L_p is only slightly less than 1.

Chapter Eight

Field-Effect Transistors

The development of the bipolar transistor, which is now sometimes called the BJT, has opened the floodgates for the development of solid-state components and the high-tech industry that accompanies it. Never again will it be possible to put a finger in the dike and hold back what can be viewed as the tsunami of technological development of the last half of the twentieth century. This ever-persistent onrush has created the industry of today and, as every politician hopes, the promise of a local industry before the next election.

The reader may be at a stage of life where it is common to wax nostalgic over the good old days before the development of the "computer on the wrist" and recall some youthful pranks. There was Father watering the lawn with the long garden hose, spraying a steady stream of water on the parched grass. This was a time for youthful mischievous activity, that of jumping up and down on the hose before it came around the house and into Father's view. How would Father ever understand these uneven eructations of water from the nozzle? Little did the youth in wildest imagination ever dream that jumping up and down and creating a change of flow in the hose would be studied later in school as a fundamental concept and that it would lead to the development of the ——— FET industry. The blanks can be filled in with several combinations of various letters, such as MOS, MIS, and IG. Is there another impish episode from the days gone by that can lead to a new high-tech industry? We shall see, as the days pass and the fruits are harvested from the seeds sown early during the reader's lifetime.

I. Field-Effect Transistor Operation

The basic idea of operation for the field-effect transistor was formulated independently by J. E. Lilienfeld in the United States and O. Heil in Germany during the period when physics was undergoing the quantum mechanics revolution at the time of the Great Depression. Patents were awarded some 20 or so years ahead of the discovery of the bipolar transistor. Their idea was simple and can be described with the following model. Imagine a semiconducting slab that has an electrode attached to each end. A third electrode is laid on top of the semiconductor, as shown in Figure 8.1. By applying a voltage to this third electrode, they felt that the electrical conductance between the outer two electrodes would be modulated. Such was the early idea that would later reach

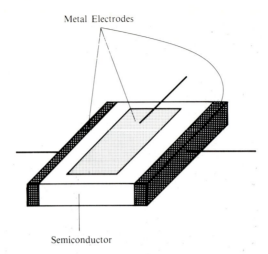

Figure 8.1   Basic field-effect transistor.

fruition in the *field-effect transistor*. At the time, its development into a practical device was inhibited by the lack of pure semiconducting materials, technological immaturity, and the period of worldwide insanity that followed the quantum-mechanics revolution.

   To understand the details of the basic operation of a FET, consider the *pnp* transistor shown in Figure 8.2. An insulator separates the three electrodes, which are given the names *source*, *gate*, and *drain*. These terms in some fashion describe, as we will see, the function of each electrode in that the gate can inhibit the flow of charged particles from the source to the drain, where the other two names are almost self-explanatory. The important region of operation for the FET occurs between the two vertical lines in Figure 8.2 (i.e., under the gate electrode). The conducting paths between the other two electrodes and these lines can be replaced with resistors of the appropriate value, which would reflect the nonzero "ohmic" connections of these regions.

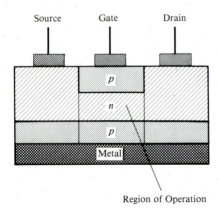

Figure 8.2   Detailed picture of a *pnp* transistor with the region of FET operation indicated.

Let us examine in detail the operation of the transistor in the region between the vertical lines, which has been expanded in Figure 8.3. In addition, the potential distribution along the *channel* between the source and the drain is also illustrated. To simplify the model, the resistances that were postulated to occur between the source and drain electrodes and this FET region $0 \le x \le L$ are set equal to zero. They could be included later as a refinement.

Certain general characteristics of the FET can be obtained from Figure 8.3. The potential ϕ_D applied to the drain that is connected to the *n*-type semiconductor is positive with respect to the *p*-type semiconductors on either side. Therefore a large depletion layer may be created at the junctions of the two semiconductors since each *pn* junction is reverse biased. If the *p* regions are very heavily doped ($N_a \gg N_d$), we find from Equation (16) of Chapter 6 that the depletion layer will reside mainly in the *n*-type semiconductor, which is sometimes called the channel. This potential ϕ_n is nonuniformly distributed in the longitudinal coordinate x and, as a first but very useful approximation, can be written as

$$\phi_n(x) = \phi_D\left(\frac{x}{L}\right) \tag{1}$$

where the potential at $x = 0$ is taken to be zero and that at $x = L$ is taken to be ϕ_D since the resistors connecting these points to the external electrodes have been removed from the model. An incremental value of the resistance of the channel is given by

$$\Delta R = \frac{\text{length}}{(\text{conductivity})(\text{cross-sectional area})}$$

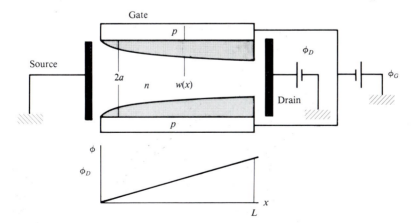

Figure 8.3   Two-dimensional representation of an FET. The third dimension b is into the paper.

or

$$\Delta R = \frac{\Delta x}{\sigma b\{2a - 2\sqrt{[2\varepsilon/(qN_d)][\phi_G + \phi_n(x)]}\}} \tag{2}$$

where the cross-sectional area of the channel between the depletion layers, which depends on position x, is given approximately by

$$b \cdot (2a - 2w(x))$$

where $w(x)$ is the depletion width, which is mainly in the n region at one of the two identical pn junctions.

At some value of voltage, the depletion widths will just touch. This location will first be at $x \approx L$ if a voltage ϕ_D is applied to the drain terminal. If $\phi_D = 0$ and only a voltage is applied to the gate terminal, the conjoining of the depletion layers will occur uniformly at all locations of x in the range $0 \leq x \leq L$. Both voltages will affect the dimension of the depletion layer that extends into the n-type semiconductor. Let us call the value of voltage that causes two depletion layers to touch in either of the two situations, the *pinch-off voltage* ϕ_{po}. It will be defined from (2) as the value that causes the resistance to $\rightarrow \infty$. Therefore,

$$2a - 2\sqrt{[2\varepsilon/(qN_d)]\phi_{po}} = 0$$

or

$$\phi_{po} = \frac{qN_d}{2\varepsilon}a^2 \tag{3}$$

In terms of the pinch-off voltage, this incremental resistance ΔR can be written as

$$\Delta R = \frac{\Delta x}{2\sigma ab\{1 - \sqrt{[\phi_G + \phi_n(x)]/\phi_{po}}\}} \tag{4}$$

The voltage change across this increment in resistance is found from

$$\Delta \phi_n = i_D \Delta R \tag{5}$$

where i_D is the current flowing in the channel through the resistance ΔR. Hence

$$\Delta \phi_n = \frac{i_D \Delta x}{2\sigma ab\{1 - \sqrt{[\phi_G + \phi_n(x)]/\phi_{po}}\}}$$

which can be rewritten as

$$\left\{1 - \sqrt{\frac{\phi_G + \phi_n(x)}{\phi_{po}}}\right\}\Delta \phi_n = \frac{i_D}{2\sigma ab}\Delta x$$

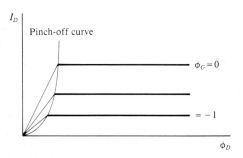

Figure 8.4   Static characteristics of a
pnp FET.

This equation can be integrated if we let $\Delta x \to dx$ and $\Delta \phi_n \to d\phi_n$. The limits of the voltage change from 0 to ϕ_D as we traverse the distance from $x = 0$ to $x = L$. Therefore, the integration becomes

$$\int_0^{\phi_D} \left\{ 1 - \sqrt{\frac{\phi_G + \phi_n(x)}{\phi_{po}}} \right\} d\phi_n = \int_0^L \frac{i_D}{2\sigma ab}\, dx$$

whose integral is

$$\phi_D - \frac{2}{3}\phi_{po}\left\{ \frac{\phi_G + \phi_D}{\phi_{po}} \right\}^{3/2} + \frac{2}{3}\phi_{po}\left\{ \frac{\phi_G}{\phi_{po}} \right\}^{3/2} = \frac{i_D L}{2\sigma ab} \qquad (6)$$

An explicit expression for the drain voltage–drain current characteristics has now been obtained. This expression, however, is valid only for voltages *less* than the value of the voltage ϕ_{po} that causes the two depletion layers to touch. For voltage values above this value, the FET is "pinched off" and the current approximately remains a constant equal to the value that it had at the transition from conduction to this new state.

In Figure 8.4, the static characteristics of the drain current versus the drain voltage for several values of gate voltage for a *pnp* FET are shown. The additional curve labeled the "pinch-off curve" will be analyzed later.

If we had included the resistances (R_S at the source region and R_D at the drain region) between the electrodes and the active region that has been analyzed previously, the external voltages would have to be modified to compensate for their contributions. The voltage at the source junction would be $\phi_{\text{source}} = I_D R_S$ if the external source electrode were grounded. Similarly, the voltage at the drain junction would be $\phi_{\text{drain}} = \phi_D - I_D R_D$ if the external drain electrode had a voltage ϕ_D applied to it. The terms R_S and R_D correspond to the values of the resistors that describe the nonohmic contacts.

Example 8.1   With the drain potential ϕ_D set equal to zero in Figure 8.3, calculate the resistance of the channel between the source and drain electrodes if there is a reverse bias applied to the gate electrode. You should assume that the *p* regions are more heavily doped than the *n* regions.

Answer: Since the p regions are more heavily doped ($N_a \gg N_d$), the depletion layer will reside mainly in the n region. The depletion width $w(x)$ will therefore be independent of the longitudinal position x, $w(x) = W$. The resistance will be

$$R = \frac{L}{(2a - 2W)b\sigma}$$

where L is the length of the channel, $2a - 2W$ is the distance between the depletion widths at the two pn junctions and $W \approx [2\varepsilon\phi_G/qN_d]^{1/2}$, b is the transverse dimension, and σ is the conductivity of the channel. In addition, we must add the resistances between the drain and source electrodes and the channel in order to compute the total resistance between the source and the drain electrodes.

II. Equivalent Circuit

Having obtained an equation relating the drain current and the gate and the drain voltages, it is possible to suggest an equivalent circuit to model the FET (sometimes called the junction FET or J-FET). This model will be particularly helpful when describing the small-signal ac response of the FET. Let us use capital letters to describe the dc bias conditions and lowercase letters to describe the ac terms. Therefore, we can write

$$i_d = I_D(\Phi_D + \phi_d, \Phi_G + \phi_g) - I_D(\Phi_D, \Phi_G)$$

Expanding the first term in a Taylor series about the point Φ_D, Φ_G, we obtain

$$i_d \approx \frac{\partial I_D}{\partial \Phi_D} \phi_d + \frac{\partial I_D}{\partial \Phi_G} \phi_g \tag{7}$$

where

$$g_d = \frac{\partial I_D}{\partial \Phi_D}, \text{ the drain or channel conductance}$$

$$g_m = \frac{\partial I_D}{\partial \Phi_G}, \text{ the transconductance or mutual conductance}$$

The g_m plays an analogous role as the α and β in modeling a bipolar junction transistor, and g_d can be thought of as either the device output admittance or the ac conductance of the channel between the source and the drain. Explicit expressions can be obtained for these terms for voltages that are below the pinch-off voltage by using (6) and performing the appropriate partial differentiations. For voltages that are greater than the pinch-off voltage, the channel conductance is zero by definition for an ideal FET. Practical FETs do have a nonzero

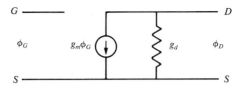

Figure 8.5 Small-signal equivalent
circuit of an FET.

conductance as some charge can migrate in the overlapping depletion layers. An
explicit expression for g_m above pinch-off is not available at this stage since (6) is
not valid in this region. Approximate values for these terms will be obtained later
when the relation between the saturated drain currents and voltages are obtained
in the section on MOSFET analysis. They can also be obtained graphically, as
will be shown in Example 8-2.

A small-signal equivalent circuit for the FET is shown in Figure 8.5. To this
model, we should incorporate capacitances across the input and the output
terminals, which will become important at higher frequencies. These elements
reflect the normal capacitances between metal conductors.

Example 8.2 Let the FET characteristics be defined with the straight-line
models shown in the figure. Find the equivalent circuit elements for Figure 8.5
at the indicated operating point. Find the pinch-off voltage if the drain voltage
is almost zero.

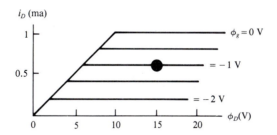

Answer: The drain conductance g_d is given by

$$g_d = \frac{(0.6 - 0.6)\ \text{mA}}{(20 - 10)\ \text{V}} = 0\ \Omega^{-1} \quad \text{or} \quad r_d \to \infty$$

The mutual conductance g_m is given by

$$g_m = \frac{(0.8 - 0.4)\ \text{mA}}{(-0.5 - [-1.5])\ \text{V}} = 4 \times 10^{-4}\ \Omega^{-1}$$

If a voltage of -2.5 V is applied to the gate electrode, the FET will be in the pinch-
off condition and the drain current will not change due to a change in the drain
voltage.

III. MOS Capacitor

To gain a deeper insight into the operation of a field-effect transistor, it is necessary to first study the energy level diagrams of a device consisting of a *p*-type semiconductor joined together with an oxide and a metal. We could later study an *n*-type semiconductor instead, but it may be easier to examine the *p*-type first, since the hole motion and the current will be in the same direction. Prior to joining these materials, their energy levels with respect to the Fermi level are as shown in Figure 8.6. Note that in the case of the metal the Fermi level is almost in the conduction band, while it is intermediate within the large gap between the valence and the conduction bands for the insulator. For a *p*-type semiconductor, it is closer to the valence band than the conduction band.

    The short horizontal lines in Figure 8.6 at the top of these energy levels indicate the vacuum levels external to the materials. This is the minimum energy required to allow the electrons to be completely free of the material. In a metal, the difference between the vacuum and the Fermi energy, $q\phi_M$ is called the *work function*. In the semiconductor, the difference between the vacuum level and the conduction band is called the *affinity χ_i*.

    If the materials are properly joined in the manufacturing process, the Fermi energies will line up since they will be in thermal equilibrium and the energy level diagram changes to that given in Figure 8.7. No bias is applied to the metal plate. The differences between the work functions of the metal, the oxide, and the semiconductor create a *contact potential* ϕ_{os} at the surface of the oxide and the semiconductor. This contact potential creates a depletion layer that is deficient in mobile holes and has a width W_d in the semiconductor, where

$$W_d = \sqrt{\frac{2\varepsilon_s \phi_{os}}{qN_a}}, \qquad \phi_G = 0 \tag{8}$$

where ε_s is the dielectric constant of the semiconductor. In this depletion layer, the charge resides on the "ionized" acceptor impurity atoms, and we will assume that there is a complete depletion of the majority carriers in the region.

    The MOS structure can now be thought of as consisting of two capacitors in series. The thicknesses of the two capacitors are W_d, which is the depth of the

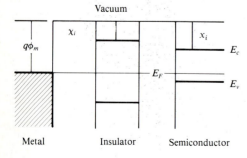

Figure 8.6   Energy level diagrams for the three components of an MOS capacitor.

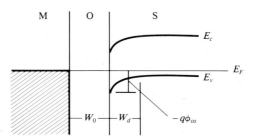

Figure 8.7 Energy level diagram for an MOS capacitor.

depletion layer, and W_O, which is the width of the oxide layer. Therefore, the total capacitance of these two capacitors, which are in series, is given by

$$\frac{1}{C} = \frac{1}{C_{\text{oxide}}} + \frac{1}{C_{\text{depletion}}} \tag{9}$$

where

$$C_{\text{oxide}} = \frac{\varepsilon_{\text{oxide}} A}{W_0}$$

and

$$C_{\text{depletion}} = \frac{\varepsilon_s A}{W_d}$$

If a voltage ϕ_G is applied to the metal plate, a portion of it will appear across the oxide and the remainder ϕ_s will affect the width of the depletion layer in the semiconductor (i.e., the structure acts as a voltage-divider circuit). The width of the depletion layer will be modified to be

$$W_d = \sqrt{\frac{2\varepsilon_s(\phi_{os} + \phi_s)}{qN_a}}, \qquad \phi_G > 0 \tag{10}$$

The energy level diagram for this MOS capacitor is shown in Figure 8.8(a). Note that the depletion width is greater than for the case when $\phi_G = 0$. This mode of operation is called the *depletion mode*. As the voltage applied to the metal is further increased, the conduction band of the semiconductor will approach the Fermi energy. Hence electrons may then be found in the conduction band just adjacent to the oxide, as shown in Figure 8.8(b). This mode is called the *inversion mode*. These electrons are very mobile and can move easily and rapidly, so the conductivity of this channel increases. If the density of electrons at the surface exceeds the density of acceptor impurities, this region has characteristics of an *n*-type semiconductor. Therefore, an externally applied potential has created a mode of operation that is similar to an *n*-type semiconductor, but in a *p*-type semiconductor. Hence this state is given the name *inversion*.

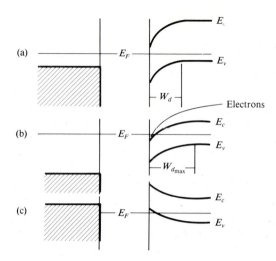

Figure 8.8 Energy levels of a biased MOS capacitor as a function of gate voltage. (a) $\phi_G > 0$. (b) $\phi_G \gg 0$. (c) $\phi_G < \phi$.

If the bias voltage applied to the metal is made negative with respect to the p-type semiconductor, mobile holes accumulate at the semiconductor–oxide interface. The conductivity of this region will also be higher, and the depletion width will decrease and eventually disappear. The energy level diagram for this *accumulation mode* is shown in Figure 8.8(c). The capacitance for the accumulation mode is due to the oxide separating the metal from the semiconductor.

  It is advantageous to summarize these effects by stating the charges that exist at the various surfaces of the MOS capacitor during the various modes of operation. The charge on the metal surface is a positive surface charge. The charge in the depletion layer consists of ionized acceptors. Conduction electrons are found in the inversion layer. The electrons stay in the inversion layer, and the depletion layer is not affected as the voltage is further increased. In the accumulation mode, it is the holes that accumulate at the surface of the metal–semiconductor interface. A common criterion for inversion is to state that the conduction band is drawn far beneath the Fermi energy. An applied bias equal to twice the Fermi energy will yield this "strong inversion."

  The charge distributions in the various regions and for the various modes are shown in Figure 8.9. For an n-type semiconductor, the signs of these charges would be reversed, and the voltage applied to the metal would have the opposite sign for the same modes.

  The MOS capacitance as a function of applied voltage on the metal is shown in Figure 8.10 for a p-type semiconductor. A similar figure could be drawn for an n-type semiconductor with only the polarity of the voltages reversed.

  Using the measured capacitance–voltage curve shown in Figure 8.10 and knowing the physical dimensions of the MOS capacitor, it is possible to estimate the density N_a of acceptors in the depletion layer whose dimension is given by (8).

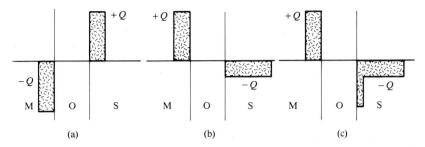

Figure 8.9 Charge distribution in an MOS capacitor. (a) Accumulation mode, (b) depletion mode, and (c) inversion mode.

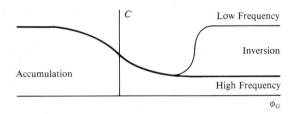

Figure 8.10 Capacitance of a MOSFET capacitor as a function of gate voltage for a *p*-type semiconductor.

This is facilitated by plotting the measured capacitance versus $[\phi_G]^{1/2}$ if it is known that the density distribution is uniform in the depletion layer.

The MOS capacitor can be used as a charge storage element or, as will be described in the next section, in a field-effect transistor configuration. This device is called a MOSFET.

IV. MOSFET

The general idea of joining a metal, an insulator, and a semiconductor in an FET device carries the acronym MISFET. Another appellation for this metal oxide–semiconductor transistor is a MOST. Since an insulator is inserted between the metal (the gate) and the semiconductor, it is also called an insulated gate FET or IGFET. The use of a *p*-type semiconductor could possibly carry the name PIGFET. The application of a reverse-biased Schottky barrier (metal–semiconductor) has the name MESFET. The names go on and on.

Finally, if the semiconductor is *silicon* and the insulator is *silicon dioxide*, the device is called a MOSFET. This is the device that is the industrial and technological giant, and hence this special name is reserved for it. As it is so important, we will examine the properties of the metal–SiO_2–Si system in detail. A typical metal that is used is aluminum.

In Figure 8.11, a cross section of a typical MOSFET is shown. It is essentially an MOS capacitor with *pn* junctions placed at either end out of the

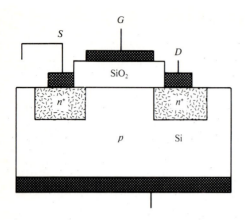

Figure 8.11 MOSFET transistor. Bias voltages are applied to the gate and drain terminals.

regions of the MOS capacitor operation. As in a normal FET (or JFET) operation, the charge carriers enter at the source and leave at the drain after passing through the gate region. The voltages relative to ground are ϕ_G and ϕ_D. The source will be assumed to be grounded, as is the backside of the device. The diode at the drain is reverse biased under normal operating conditions. The drain current I_D for a p-type bulk device is taken to be positive if it is flowing into the drain terminal.

Let us first examine the operation of the MOSFET under conditions when the potential applied to the drain ϕ_D is equal to zero. If the voltage that is applied to the gate $\phi_G < \phi_T$, where ϕ_T is the threshold voltage, it will cause the MOS capacitor region to go into the inversion mode; then the region under the gate will contain an excess or a deficit of holes, but no electrons. Hence the conducting path between the source and the drain will be poor. However, for $\phi_G > \phi_T$, negative charge will be induced in the oxide–semiconductor interface. The minority carrier electrons in the inversion layer are very mobile and will provide a high-conductivity path between the source and drain. Majority electrons from the heavily doped drain region will be easily transported through the channel region to the heavily doped source region. The MOSFET is sometimes designated as being a majority carrier device.

If now the drain voltage is slowly increased in equal increments, then a current I_D will flow through this *inversion channel*. This is shown in Figure 8.12. This current may be on the order of 1 billion times the current flow when no bias voltage is applied to the gate terminal that created the inversion layer. Although the conductivity is high, it is not infinite, and there will be a voltage drop in this channel that depends on position. This voltage drop will tend to locally negate the effect of the gate voltage that caused the inversion channel to form originally, and the dimension of the inversion channel will decrease as we move from the source to the drain. A pinch off will occur at a critical value of drain voltage $\phi_D = \phi_{D(sat)}$. For values of the drain potential $\phi_D > \phi_{D(sat)}$, the pinched-off portion increases in extent from a point to a length ΔL. For $\Delta L \ll L$, where L is the

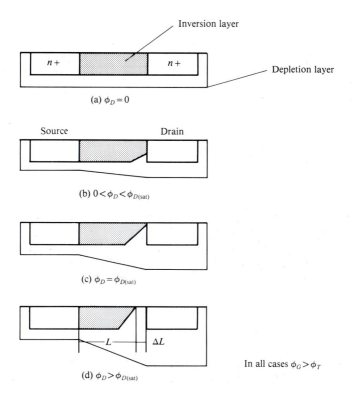

Figure 8.12 Visualization of various phases of MOSFET operation for various values of drain voltage.

length of the channel, most of the voltage drop in excess of $\phi_{D(\text{sat})}$ will appear in the region ΔL, and the current I_D will saturate at the value found when $\phi_D = \phi_{D(\text{sat})}$. The current–voltage characteristics for a MOSFET are shown in Figure 8.13. The electrons are the channel carriers in this *p*-type MOS device. This is called an *n-channel* device. Similarly, if the carriers were holes in an *n*-type MOS device, it would be called a *p-channel* device. In either case, the transistor turns on or starts to carry current at the onset of inversion.

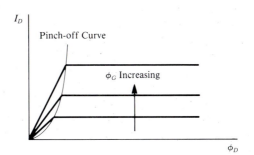

Figure 8.13 Voltage–current characteristics of a MOSFET.

V. MOSFET Analysis

As the MOSFET is so important in the semiconductor business, we will delve a little deeper into the mathematical description for the device. Many of the relations have been derived previously, but it is convenient to have all the germane expressions in one section.

First, the electrons are moving in the n channel in the p-type semiconductor due to the inversion conditions. These electrons will be accelerated in two directions. First, they will be accelerated toward the SiO_2 surface due to the voltage that is applied to the gate terminal. Second, they will be accelerated in the general direction toward the drain from the source due to the positive potential applied to the drain terminal. In this period of acceleration, the electron will possibly encounter the surface, a lattice, or an ionized impurity. It will collide and scatter from these objects, and we will assume that it starts with zero velocity after each collision and is accelerated again. On the average, there will be an overall drift of these electrons from the source to the drain. Hence it is reasonable to define the mobility μ_n of the electrons as

$$\mu_n(y) = -\frac{q}{Q_n(y)} \int_0^{X(y)} \mu_n(x, y) n(x, y)\, dx \tag{11}$$

where y is the direction along the channel, x is the direction into the inversion layer whose depth is X, $n(x, y)$ is the local density, and

$$Q_n = -q \int_0^{X(y)} n(x, y)\, dx \tag{12}$$

is the charge stored in the channel at a particular value of y. It is assumed that the channel length is much greater than the depth X.

If the voltages satisfy the relations

$$\phi_G > \phi_T, \qquad 0 \le \phi_D \le \phi_{D(\text{sat})}$$

the current density is given by

$$j_n = q\mu_n nE + qD_n \frac{\partial n}{\partial y} \tag{13}$$

where we have assumed that the dominant direction of flow is in the y direction. It is not unreasonable to neglect the diffusion term with respect to the conduction term and make the electrostatic approximation for the electric field. Therefore,

$$j_n \approx -q\mu_n n \frac{\partial \phi}{\partial y} \tag{14}$$

The current to the drain is given by

$$I_D = -\iint j_n \, dx \, dz = -\Delta z \int j_n \, dx = -\Delta z \frac{\partial \phi}{\partial y} \left(-q \int_0^X \mu_n(x, y) n(x, y) \, dx \right)$$

(15)

where Δz is the transverse width of the channel. If the mobility μ_n is replaced by its average value over the channel $\langle \mu_n \rangle$, the integral of (15) from $y = 0$ to $y = L$ can be written as

$$\int_0^L I_D \, dy = I_D L = -\frac{\Delta z \langle \mu_n \rangle}{L} \int_0^{\phi_D} Q_n \, d\phi$$

(16)

To complete the derivation of the MOSFET behavior, we must know the dependence of the charge Q_n on the potential variation along the inversion channel. The simplest argument is to consider the region as a parallel plate capacitor with charge being introduced in from either end of the channel. The total charge that is stored in this capacitor is Q_n, where

$$Q_n = -\left(Q \bigg|_{\text{inv}|_{\phi_G \geq \phi_T}} - Q \bigg|_{\text{inv}|_{\phi_G = \phi_T}} \right) \approx -C_0(\phi_G - \phi_T)$$

(17)

In this equation, ϕ_G is the potential applied to the gate and ϕ_T is the gate voltage that causes the transition from depletion to inversion. Note that the potential in the capacitor changes with position y also, this being reflected in how the drain voltage ϕ_D is distributed in space. Therefore, (17) should be replaced by

$$Q_n(y) = -C_0(\phi_G - \phi_T - \phi)$$

(18)

where $\phi = \phi(y)$.

The integral of (16) becomes

$$I_D = \frac{\Delta z \langle \mu_n \rangle C_0}{L} \left[(\phi_G - \phi_T)\phi_D - \frac{\phi_D^2}{2} \right]$$

(19)

which is valid for $0 \leq \phi_D \leq \phi_{D(\text{sat})}$ and $\phi_G \geq \phi_T$. Expressions for the saturated currents and voltages can be found from (19). This can be effected by noting that at saturation

$$Q_n(L) \to 0 \qquad \text{for } \phi_D \to \phi_{D(\text{sat})}$$

From (18), we write

$$Q_n(L) = -C_0(\phi_G - \phi_T - \phi_{D(\text{sat})}) = 0$$

This leads to

$$\phi_{D(\text{sat})} = \phi_G - \phi_T \tag{20}$$

and

$$I_{D(\text{sat})} = \frac{\Delta z \langle \mu_n \rangle C_0}{L} (\phi_G - \phi_T)^2 \tag{21}$$

This is the "square law" relation of saturated drain current as a function of gate voltage greater than the critical value. This is indicated in Figure 8.13 as the pinch-off curve in the voltage–current characteristics of a MOSFET.

At this point, it is useful to summarize the properties of the region under the gate electrode under conditions of saturation. The number of charge carriers that pass from the source to the drain remains the same, as does the current that flows from the source to the drain. The potential at a point under the gate electrode will remain at the saturated value $\phi_{D(\text{sat})}$, except for the variation along the channel due to the potential applied to the drain.

An alternative approach can be used to derive the expression given in (20). That is to extend the region of validity of (19) to include values of drain potential ϕ_D that are greater than the saturated value $\phi_{D(\text{sat})}$. Since the drain current has saturated, the derivative $\partial I_D / \partial \phi_D$ will equal zero. Differentiating (19) and setting it equal to zero will yield (20). We should be careful about employing this approach in a cavalier fashion without considering the consequences of using an expression whose range of validity is limited.

Example 8.3   Find the potential distribution in the channel region of a MOSFET when the current just saturates at $I_D = I_{D(\text{sat})}$.

Answer: From (16), we write

$$I_{D(\text{sat})} y = -\frac{\Delta z \langle \mu_n \rangle}{L} \int_0^{\phi(y)} Q_n \, d\phi$$

where Q_n is defined in (18). The integral leads to

$$I_{D(\text{sat})} y = \frac{\Delta z \langle \mu_n \rangle C_0}{L} \left\{ (\phi_G - \phi_T)\phi(y) - \frac{\phi(y)^2}{2} \right\}$$

Solving for $\phi(y)$ with $\phi_{D(\text{sat})} = \phi_G - \phi_T$, we write

$$\phi(y) = \frac{\Delta z \langle \mu_n \rangle C_0}{L} \phi_{D(\text{sat})} \left\{ 1 \pm \sqrt{1 - \frac{2 I_D y}{\left(\frac{\Delta z \langle \mu_n \rangle C_0}{L} \phi_{D(\text{sat})} \right)^2}} \right\}$$

With the relation for the saturation current given in (21), it is possible to derive expressions for the equivalent circuit shown in Figure 8.5. The mutual conductance is given by

$$g_m = \frac{\partial I_{D(\text{sat})}}{\partial \phi_G} = \frac{2 \, \Delta z \langle \mu_n \rangle C_0}{L} (\phi_G - \phi_T)$$

and the channel conductance g_d is given approximately by

$$g_d(\phi_D \to 0) = \frac{\partial I_{D(\text{sat})}}{\partial \phi_D} = \frac{2 \, \Delta z \langle \mu_n \rangle C_0}{L} (\phi_G - \phi_T)$$

where (20) has been freely employed.

The square-law relation given in (21) appears to be quite reasonable. There is, however, a major flaw in the analysis, which can be corrected. A tacit assumption has been made in the derivation. The analysis has assumed that the depletion width $W(y)$ remains the same for all values of y. However, from Figure 8.12 it is noted that the depletion width depends on the location y. If this is included, (18) is modified to

$$Q_n(y) = -C_0(\phi_G - \phi_T - \phi) + qN_aW_T\left(\frac{W(y)}{W_T} - 1\right) \tag{22}$$

where W_T and $W(y)$ are the depletion width with $\phi = \phi_T$ and the depletion width along the channel, respectively. The ensuing algebra is better left to others. We find that a correction term is obtained that reduces the values of $I_{D(\text{sat})}$ and $\phi_{D(\text{sat})}$ from those given with the square-law theory. The basic square-law theory explains the operation, and its accuracy improves with decreasing substrate density.

VI. Transient MOSFET Analysis

In the description of the MOSFET presented previously, a steady-state analysis was employed to determine the behavior of the charge in the inversion channel that connects the source and drain regions. There are important cases where the transient behavior should be examined, in particular, cases where high-speed switching is desired, as in modern digital circuits.

The equation that will govern the temporal and spatial behavior of the electrons in the inversion layer of the MOSFET is the equation of continuity, which is given as

$$q\frac{\partial n}{\partial t} + \frac{\partial j_n}{\partial y} = 0 \tag{23}$$

where the current j_n is defined in (13) as

$$j_n = q\mu_n nE + qD_n \frac{\partial n}{\partial y} \tag{13}$$

Note that only the current that flows between the drain and the source is being considered. The electric field is found from (18) by just adding the contribution of the field due to the potential applied between the source and the drain, as follows:

$$\phi(y) \approx \gamma\left\{(\phi_G - \phi_T) + (\phi_D - \phi_S)\frac{y}{L} + \frac{qn(y)}{C_0}\right\} \tag{24}$$

where $Q_n(y) = qn(y)$, γ is a constant, and a linear variation of potential between the source and the drain is assumed. The electric field is given from $E = -\partial\phi/\partial y$, from which we compute

$$E = -\gamma\left\{\frac{\phi_D - \phi_S}{L} + \frac{q}{C_0}\frac{\partial n(y)}{\partial y}\right\} \tag{25}$$

The first term is the drift field caused by the difference in potentials between the source and the drain, and the second term is the self-induced field due to the nonuniform charge distribution. Therefore, (23) becomes

$$q\frac{\partial n}{\partial t} + \frac{\partial j_n}{\partial y} = 0$$

$$q\frac{\partial n}{\partial t} + \frac{\partial[q\mu_n nE + qD_n(\partial n/\partial y)]}{\partial y} = 0$$

$$q\frac{\partial n}{\partial t} + \frac{\partial[-q\mu_n n\gamma\{[(\phi_D - \phi_S)/L] + (q/C_0)(\partial n/\partial y)\} + qD_n(\partial n/\partial y)]}{\partial y} = 0$$

$$\tag{26}$$

where $n = n(y, t)$. This can be rewritten as

$$\frac{\partial n}{\partial t} - \gamma\mu_n\left\{\frac{\phi_D - \phi_S}{L}\frac{\partial n}{\partial y} - \frac{\gamma\mu_n}{C_0}\frac{\partial\{n(\partial n/\partial y)\}}{\partial y}\right\} + D_n\frac{\partial^2 n}{\partial y^2} = 0 \tag{27}$$

The first term of (27) corresponds to the temporal variation of the charge density, the second term is due to the drift created by the difference in potentials between the drain and the source terminals, the third term results from the nonuniform charge distribution, and the fourth term is due to diffusion. Equation (27) is nonlinear and our first inclination is to look at a numerical solution. However, an analytical solution will be obtained, as is shown next.

Let us first transform (27) into the "drift frame" where

$$\phi = y + \gamma\mu_n\frac{\phi_D - \phi_S}{L}t \tag{28a}$$

$$\tau = t \tag{28b}$$

From the chain rule, we calculate

$$\frac{\partial}{\partial t} = \frac{\partial}{\partial \tau}\left(\frac{\partial \tau}{\partial t}\right) + \frac{\partial}{\partial \phi}\left(\frac{\partial \phi}{\partial t}\right) = \frac{\partial}{\partial \tau} + \gamma\mu_n\frac{\phi_D - \phi_S}{L}\frac{\partial}{\partial \phi} \tag{29a}$$

$$\frac{\partial}{\partial y} = \frac{\partial}{\partial \tau}\left(\frac{\partial \tau}{\partial y}\right) + \frac{\partial}{\partial \phi}\left(\frac{\partial \phi}{\partial y}\right) = \frac{\partial}{\partial \phi} \tag{29b}$$

Equation (27) can be written in these variables as

$$\frac{\partial n}{\partial \tau} - \frac{\gamma\mu_n}{C_0}\frac{\partial\{n(\partial n/\partial \phi)\}}{\partial \phi} + D_n\frac{\partial^2 n}{\partial y^2} = 0 \tag{30}$$

or

$$\frac{\partial n}{\partial \tau} - \left[\frac{\gamma\mu_n}{C_0}n - D_n\right]\frac{\partial^2 n}{\partial \phi^2} - \frac{\gamma\mu_n}{C_0}\left(\frac{\partial n}{\partial \phi}\right)^2 = 0$$

Writing (30) in this latter form allows us to recognize that (30) can be expressed as

$$\frac{\partial \eta}{\partial \tau} - \frac{\partial[\eta(\partial n/\partial \phi)]}{\partial \phi} = 0 \tag{31}$$

where

$$\eta = n - \frac{D_n}{\gamma\mu_n/C_0}$$

Equation (31) is a *nonlinear* diffusion equation since the diffusion coefficient depends on the amplitude η! Writing (27) in this more compact form allows us to more easily recognize the technique for finding a solution of the equation.

To obtain a solution of (31), let us use the same technique that was used to obtain a solution of the linear diffusion equation given in Chapter 3, Section III. As the results of this procedure are so important for solving diffusion-type equations, which are found so often in semiconductor calculations, a mathematical introduction to the technique is included in Appendix A with particular emphasis on finding the solution of (31). Suffice it to say at this point in our attempt to gain an understanding of a MOSFET that we will just use the "old-world" approach of stating, "Let us use the *Ansatz* that the variables should be combined in the form

$$\Omega = \frac{\phi}{\tau^{1/3}} \tag{32}$$

and

$$\Phi(\Omega) = \frac{\eta}{\tau^{-1/3}} \tag{33}$$

Substituting (32) and (33), we again obtain an ordinary differential equation, which can be written as

$$\frac{1}{3}\frac{d(\Omega\Phi)}{d\Omega} + \frac{d[\Phi(d\Phi/d\Omega)]}{d\Omega} = 0 \tag{34}$$

To make the transformation given in (32) and (33) meaningful, the boundary and initial conditions, of which there are three in (31), must *consolidate* into the two that are required for solving (34). We find that $\eta(\phi \to \infty, \tau) \to 0$ and $\eta(\phi, \tau \to 0^+) \to 0$ consolidate to $\Phi(\Omega \to \infty) \to 0$. These two physical boundary conditions tell that the number of particles at very large distances is vanishingly small and that the particles cannot move a finite distance in a nonzero time interval. Hence the first integral of (34) with the ensuing constant of integration set equal to zero leads to

$$\frac{1}{3}\Omega\Phi + \Phi\frac{d\Phi}{d\Omega} = 0 \tag{35}$$

The integral of (35) is

$$\tfrac{1}{6}\Omega^2 + \Phi = K \tag{36}$$

where K is a constant of integration.

Rewriting (36), we obtain

$$\Phi = K - \tfrac{1}{6}\Omega^2 \tag{37}$$

The constant K will be determined as follows. The particles cannot have an infinite velocity. Therefore, for some value of Ω given by (32), no particles will be found. Let us call this critical value Ω_0. Therefore, from (37) we write

$$0 = K - \tfrac{1}{6}\Omega_0^2 \tag{38}$$

Therefore,

$$\Phi = \frac{\Omega_0^2}{6}\left[1 - \left(\frac{\Omega}{\Omega_0}\right)^2\right] \tag{39}$$

The solution of this nonlinear diffusion equation has been obtained. We can transform it back into the coordinates of the MOSFET as follows:

$$\eta = \frac{\Phi}{\tau^{1/3}}$$

$$n = \frac{D_n}{\gamma\mu_n/C_0} + \eta = \frac{D_n}{\gamma\mu_n/C_0} + \frac{\Phi}{\tau^{1/3}} = \frac{D_n}{\gamma\mu_n/C_0} + \frac{1}{\tau^{1/3}}\left\{\frac{\Omega_0^2}{6}\left[1 - \left(\frac{\Omega}{\Omega_0}\right)^2\right]\right\}$$

or finally

$$n = \frac{D_n}{\gamma\mu_n/C_0} + \frac{\Omega_0^2}{6t^{1/3}}\left[1 - \frac{\{y + \gamma\mu_n[(\phi_D - \phi_S)/L]t\}^2}{6\Omega_0^2 t^{2/3}}\right] \tag{40}$$

The analytical solution given in (40) predicts many of the features that could be found in a direct numerical solution of (27) in that (1) the maximum of the front moves with a constant velocity given by $\gamma\mu_n[(\phi_D - \phi_S)/L]$, (2) the front diffuses in time, and (3) the peak decays in time as $(1/t^{1/3})$. This solution is the solution for an "impulse" initial condition, although other initial conditions may asymptotically approach it. A normalized charge behavior in space at various times is shown in Figure 8.14. Note that the area under each of the curves remains the same as time increases, which indicates that the charge is conserved as it passes down the line.

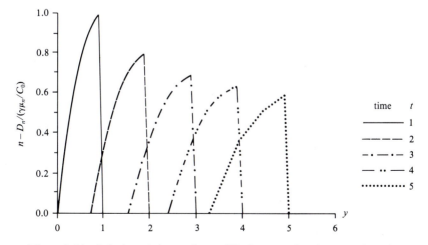

Figure 8.14 Solution of the nonlinear diffusion equation due to an impulse of charge at $y = 0$ and $t = 0$.

VII. Threshold Voltage for MOSFET

In the analysis of the MOSFET, we encountered a voltage called the *threshold voltage* ϕ_T, which was defined as the critical voltage at which an inversion layer would be formed under the gate electrode in the semiconductor. The reader is probably wondering if any external parameters can be altered in order to change the actual value of ϕ_T. The most obvious method of having some control of this voltage is to change the thickness of the insulating layer that separates the metal from the semiconductor, since the voltage drop that appears across the insulator will not alter the depletion width of the channel.

    Another technique to change the value of ϕ_T is to change the substrate doping or to actually implant ions in the substrate, possibly boron or phosphorus ions, in the desired channel location during manufacture. The use of boron will cause a positive shift in ϕ_T, and the use of phosphorus will cause a negative shift. Also, the substrate surface orientation can be altered or the material used as the gate could be changed, say from aluminum to polysilicon.

    The techniques outlined can be used only in the manufacturing process of the MOSFET. But is it possible to change the threshold voltage after the manufacture? The answer is that it can be modified by *back biasing* the MOSFET. In this condition, the transistor is constructed on an additional metal plate during manufacture such that we have a metal oxide–semiconductor– metal configuration. Through external biasing, ($\phi_{bias} < 0$ for an n-channel device and $\phi_{bias} > 0$ for a p-channel device), we find that back biasing increases ϕ_T. It cannot decrease it.

VIII. Conclusion

The field-effect transistor has led to several new industries and an understanding of its operation follows from the traversal of charge in an inversion layer or in a channel that is sandwiched between two similar-type semiconductors. Both steady-state and transient charge movements have been described. It has been possible to derive expressions relating the output current to the voltage. Certain comments have been added on the modifications of the expected operation due to manufacturing techniques or additional biasing voltages.

PROBLEMS

1. Find the resistance between $0 \leq x \leq L$ of the channel of a MOSFET if the two electrodes are divided into two equal parts and the drain potential is applied to the infinitesimally thin grid at $x = L/2$. The grid is $\approx 100\%$ transparent. Also, the magnitude of ϕ_D is equal to ϕ_G. The third dimension of the electrodes is b.

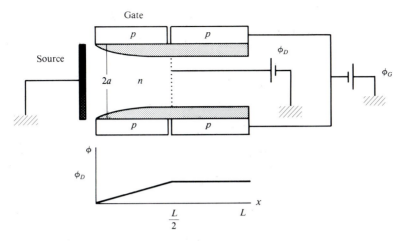

2. Calculate the pinch-off voltage for the FET described in Problem 1 if $\phi_G > \phi_D$ and if $\phi_D > \phi_G$. Discuss your results.

3. For an ideal MOS capacitor, the applied potential ϕ must be equal to twice the Fermi potential ϕ_F for the capacitor to pass into the inversion mode. If the semiconductor is p-type silicon with $N_a = 10^{15}$ cm^{-3}, calculate the value of this voltage at room temperature. Find the depletion width corresponding to this potential. The relative dielectric constant of silicon is 11.8.

4. It is possible to model the FET characteristics with the expression

$$I_D \approx I_{D0}\left(\frac{\phi_{po} + \phi_G}{\phi_{po}}\right)\tanh\left(\frac{\phi_D}{\phi_{D0}}\right) \quad \text{for} \quad \frac{\phi_D}{\phi_{D0}} \geq 1 \quad \text{and} \quad -\phi_{po} \leq \phi_G \leq 0$$

Using this description, obtain the equivalent circuit elements for the FET shown in Figure 8.5 for values of potential ϕ that are above and beneath the pinch-off conditions. Also plot the FET characteristics. You should increment ϕ_G/ϕ_{po} in reasonably small values.

5. Draw the energy level diagram for an MOS capacitor if an n-type semiconductor was used for the three regions of operation based on the different biases applied to the gate electrode.

6. Sketch the charge distribution for the three modes of operation of an MOS capacitor with an n-type semiconductor.

7. Find the electric field distribution in the channel region of a MOSFET.

8. Find the combination of independent and dependent variables that will allow the partial differential equation $\partial n/\partial t - \partial[n^2\,\partial n/\partial x]/\partial x = 0$ to be transformed into an ordinary differential equation and satisfy the condition that $n(x \to \infty, t) \to 0$.

9. Other solutions than that given here for (31) exist that will satisfy different boundary conditions. Show that $\eta = H(\phi/\tau^{1/3})$ will satisfy (31). This function will also satisfy $\eta(\phi = 0, \tau) = $ constant, which would correspond to an infinite reservoir at $\phi = 0$. Sketch the expected solution.

10. Discuss the frequency limitations of a MOSFET in terms of charge transport.

Chapter Nine

Integrated Circuits

The devices that have arisen since the quantum mechanics revolution have been numerous and, more importantly, small. Following the technology that permitted the production of the alphabet soup of transistors during the middle third of the twentieth century, another revolution in technology was forthcoming during the last third of the century. The basic laws of nature would not be brought into question as they had been earlier, but the ingenuity of the engineer and the vision of the entrepreneur would be tested as they had never been before. Solid-state physics had provided a foundation for the development of the transistor, but nature had imposed a limitation on the minimum time interval ΔT at which an effect could be felt due to some cause. Einstein had shown that nothing could go faster than the speed of light c. Hence we had an ultimate lowest limit for this time of $\Delta T = \Delta X/c$, where c is the velocity of light and ΔX is the distance between the cause and the resulting effect. To minimize this time interval ΔT, the separation ΔX had to be minimized. Hence the tube gave way to the transistor, each with their external components consisting of resistors, capacitors, and inductors that were connected with wires and soldered connections. And all these are now giving way to the integrated circuit whose size marches to the tune of the Disney song, "It's a small world, after all... It's a small, small world!"

In addition to increasing the speed of the ultimate use of an electronic component, the integrated circuit has allowed a higher density of components to reside with the angels on the head of a pin. As the circuit density can be very high, alternate electronic paths for a particular operation can be designed into the "chip" such that the chip can compensate for failures. The integrated circuit has also increased the reliability of and decreased the cost in the manufacturing process, since the design can be more accurately and cheaply manufactured by machine than with human fingers. There is reproducibility on a large scale. If the learned doctors of medicine could some day "engineer the genes" properly, perhaps our fingers will become smaller and the "computer on the wrist" will govern our lives.

In this chapter, we present some ideas concerning the manufacture of integrated circuits and define the terminology that is used in VLSI circuit design. This will give the reader a taste of the technology involved in the industry.

I. Basic Ideas

We usually do not manufacture one integrated circuit at a time; hundreds to thousands of these circuits are manufactured at a time on a single silicon wafer. The process of manufacturing this large number of identical circuits is called *batch fabrication.* Since we obtain a large number of highly reliable circuits at a time, the cost per circuit is low although the cost for the entire procedure may not be small. The procedure for manufacturing a circuit containing 1000 transistors is almost the same as for a circuit with 100 transistors, so the ultimate cost of a chip is relatively insensitive to the number of elements that have been designed into the circuit. This increases the flexibility that the circuit designer has in the final design, since redundant elements can be added and we are not limited to the "minimum number" criterion that plagued the older designs.

   In an IC, as the integrated circuit is commonly called, the size can be minimized such that the time delay $\Delta T = \Delta X/c$ can become extremely short. Modern industry has been able to reduce the dimensions of a complicated circuit to be on the order of 1 centimeter or less. Hence a signal will traverse across the entire circuit in a time ΔT, where

$$\Delta T = \frac{\text{circuit size}}{c} \approx \frac{10^{-2} \text{ m}}{3 \times 10^8 \text{ m/s}} \approx 3 \times 10^{-11} \text{ sec}$$

For distances this small, the size of a particle of dust, which may be $\frac{1}{2}$ micrometer in diameter, could have very deleterious effects since this dimension may be on the order of a circuit component. It is frequently advantageous to actually manufacture many smaller circuits rather than one large one, since dust could spell disaster. Better three smaller circuits that work and one that does not work than one large circuit that does not work. Is there an ultimate shortest time interval for a solid-state device? Yes, and it is on the order of the interatomic spacing:

$$\Delta T = \frac{2 \times \text{atomic dimension}}{\text{velocity of light}}$$

If we let the atomic dimension be the Bohr radius for a hydrogen atom (0.529172×10^{-10} m), this time is approximately 3×10^{-19} seconds.

   There are several ways of categorizing the IC, which are based on their use or their method of manufacture. They can be classified as *linear* or *digital*, which specifies their application, and *monolithic* or *hybrid*, which arises from their manufacturing process.

   Linear circuits include common amplifiers, operational amplifiers, and analog circuits as examples. Digital circuits comprise the largest application of the IC manufacturing process since the digital circuit requires just an electronic "switch" that is either on or off. Therefore, the manufacturing requirements are not as stringent for digital applications as they are for analog applications. Fortunately, computers are designed with just zeros and ones with no intermediate "one-halves." The manufacture of the transistor is not dimensional tolerance limited, although it is more difficult to construct a resistor or capacitor that does depend on physical dimensions.

A monolithic circuit contains an entire circuit on a single piece of semiconductor, which is usually silicon. A hybrid circuit may include one or more monolithic circuits, transistors, resistors, capacitors, or other elements that are bonded to an insulating substrate. The former allows for batch manufacturing and duplication to a high degree, while the latter provides the opportunity of including precision external resistors and capacitors. For small-volume production, the hybrid process may be more economical.

One final classification remains, that based on the procedure employed to fabricate and connect external resistors and capacitors to monolithic circuits. Either a *thick-film* or a *thin-film* process is employed. No firm definition separates the two thicknesses for ICs. A typical definition for a thin film is that it is 0.1 to 0.5 μm thick while a thick-film is 25 μm thick. In the thick-film process, resistors and connections are "printed" on a ceramic substrate with, say, conducting pastes consisting of metallic powders in organic binders, which are then passed through a silk screen. The resulting pattern is baked in an oven. Precise values of resistors can be obtained by mechanically trimming the printed pattern, possibly with a precisely controlled laser. Small ceramic-chip capacitors are bonded into the circuit along with monolithic circuits and individual transistors. This additional step will increase the cost.

Thin-film connecting patterns and resistors are usually vacuum deposited on a glass substrate. Resistors are made of a resistive metal such as tantalum that is sputtered, and the conductors are either aluminum or gold. A common procedure is to cover the entire surface with a metal and then remove selected portions of it using photolithographic techniques until the desired pattern is achieved. Capacitors are made by depositing an insulating layer between two conducting layers.

In our discussion of the manufacturing procedure, no mention was made of the third passive component, the inductor. There is a reason for this. They are difficult to integrate into the small substrates and thus the designer should try to avoid incorporating them into a design.

II. Fabrication

The procedure for growing a pure silicon crystal used in integrated circuits usually follows the Czochralski technique. In this technique, we start with a relatively pure form of sand (SiO_2), which is then placed in an oven with various forms of carbon (coal, coke, and wood chips). The overall chemical reaction is

$$SiC \text{ (solid)} + SiO_2 \text{ (solid)} \longrightarrow Si \text{ (solid)} + SiO \text{ (gas)} + CO \text{ (gas)}$$

The purity of the silicon is about 98% at this stage. The silicon is pulverized and treated with hydrogen chloride at a temperature of 300 K to form trichlorosilane:

$$Si \text{ (solid)} + 3HCl \text{ (gas)} \longrightarrow SiHCl_3 \text{ (gas)} + H_2 \text{ (gas)}$$

Trichlorosilane is a liquid at room temperature and it can be purified through distillation of the liquid. Finally, pure silicon is obtained through a hydrogen reduction process:

$$SiHCl_3 \text{ (gas)} + H_2 \text{ (gas)} \longrightarrow Si \text{ (solid)} + 3HCl \text{ (gas)}$$

This latter process takes place in a reactor containing a resistance-heated silicon rod on which the pure silicon is deposited. The resulting silicon has an impurity content that is measured in the parts-per-billion range. This is the raw material used for growing a single-crystal silicon.

A typical Czochralski crystal puller is shown in Figure 9.1. The apparatus has three main components: (1) a furnace, which includes a fused-silica (SiO_2) crucible, a graphite holder, a rotation mechanism, and a heating element; (2) a crystal-pulling mechanism, which includes a seed holder that can rotate in synchronism with the crucible; and (3) an ambient control system, which includes a gas source such as argon.

The crystal is grown by inserting the tip of the seed (a pure silicon crystal) into the silicon melt and slowly withdrawing it. A progressive freezing occurs at the solid–liquid interface. In the process of growing the crystal, a known amount of dopant can be added to the melt to obtain the desired doping concentration in the grown crystal. For silicon, boron and phosphorus are the most common dopants for *p*- and *n*-type materials.

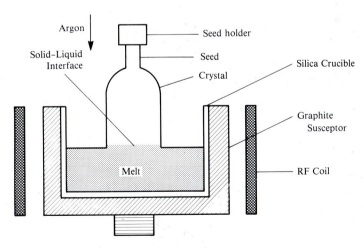

Figure 9.1   Czochralski crystal puller. The seed and seed holder rotate counterclockwise, while the graphite susceptor rotates clockwise.

After the crystal is grown, the ingot as it is now called is shaped by first removing the seed end and the opposite end, which was the last to solidify. A surface is then ground so the diameter of the material and the crystal orientation can be defined. Then one or more flat surfaces are ground along the length of the ingot so that automatic processing equipment will have a reference point to position the resulting wafer, the wafer being a thin slice (0.5 to 0.7 mm thick) that has been sliced with a diamond saw. After slicing, the wafer is lapped with a mixture of Al_2O_3 and glycerine to produce a flatness on the order of 2 μm. The possibility exists that the crystal may not be perfect. For example, it is finite in size and hence surface atoms will not have complete bonding, point defects may exist, and so on. We will not dwell on these limitations here other than to note their existence.

Starting with a silicon substrate, it is possible to grow a layer of silicon dioxide on it due to thermal oxidation. The chemical reactions that can be employed include

$$Si \text{ (solid)} + O_2 \text{ (gas)} \longrightarrow SiO_2 \text{ (solid)}$$

and

$$Si \text{ (solid)} + 2H_2O \text{ (gas)} \longrightarrow SiO_2 \text{ (solid)} + 2H_2 \text{ (gas)}$$

The SiO_2 interface moves into the silicon during the oxidation process. This creates a fresh interface region, with the surface contamination on the original silicon ending up on the oxide surface. There are several other chemical means of depositing a SiO_2 layer on the silicon substrate also.

Several techniques exist for depositing the metal surface on the SiO_2 layer. These include various evaporation techniques using ovens or electron beams to cause the evaporation, sputtering techniques, and chemical vapor deposition.

In addition to adding dopants into the melt during the process of growing the crystal, impurity atoms can be added by either diffusing them into the silicon or implanting them with a high-velocity ion beam.

The removal of controlled regions of SiO_2, say with an ion beam, an electron beam, or a laser, will expose the semiconductor. The ion beam can have a higher resolution than the electron beam due to its heavier mass creating less scatter. Chemical etching is also widely used in the industry. Metal contacts can then be bonded to the exposed semiconductor region.

The brief discussion here summarizes in a few words the technological advances of over 30 years by "the best and the brightest," whose number is in the thousands. This discussion is included only to point the reader in the "high-tech" direction.

III. Passive Components

Having just indicated that it is possible to manufacture semiconductor–insulator wafers, we might wonder if it is possible to create electronic circuit elements. This is possible and several of these elements will be described here.

To create a resistor, we create a window in the silicon insulator and implant or diffuse an impurity of the opposite conductivity into the wafer, as shown in Figure 9.2. In this case, a *p*-type material of thickness Δx is introduced into the *n*-type region. Let ΔW be the width of the channel. Note that the electrodes at each end are attached at the exposed regions, whose dimensions are $3\Delta W \times 3\Delta W$. The conductance G_{bar} between these two large regions is given by

$$G_{\text{bar}} \approx \frac{q\mu_p p \,\Delta W \,\Delta x}{L} \tag{1}$$

where we have assumed that the conductivity $\sigma = q\mu_p p$ is a constant in the channel and that it is much higher than in the underpinning semiconductor. This assumption may be quite crude since it was required to *diffuse* an impurity into silicon. From your understanding of the diffusion equation (Chapter 3, Section III), you are aware that we should assume that the profile of the diffused impurities should have a Gaussian, an integral of a Gaussian (called an erfc function, which stands for the complementary error function), or possibly an exponential profile if steady-state conditions were examined. The constant value that is assumed would correspond to a spatial average of these profiles.

If $L = n\,\Delta W$ such that there are *n* "squares" in this region, the resistance of this region can be written as

$$R_{\text{bar}} = \frac{1}{G_{\text{bar}}} = \frac{n}{\sigma \,\Delta x} = \frac{n}{g} \tag{2}$$

where $1/g$ is the resistance per square. The resistance of each electrode region contributes approximately 0.65 of a square. Therefore, the total resistance is

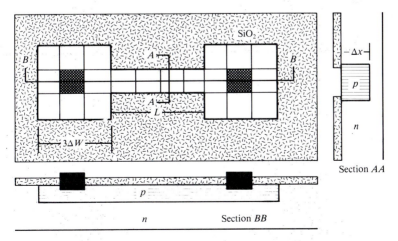

Figure 9.2   Integrated-circuit resistor. Each square is $\Delta W \times \Delta W$ in size with a thickness Δx.

equal to

$$R = R_{\text{bar}} + 2R_{\text{electrode}}$$

$$= \frac{n + 1.3}{g} \tag{3}$$

There are two techniques to construct capacitors. The first is to construct an MOS capacitor where the metal is one electrode and a localized very heavily doped semiconductor acts as the other electrode. The oxide region acts as the intervening dielectric. The capacitance per unit area is

$$C = \frac{\varepsilon_{\text{ox}}}{d} \tag{4}$$

where ε_{ox} is the dielectric constant of the oxide and d is the separation between the two electrodes. The second technique is to use a reverse-biased pn junction as a capacitor. This junction forms part of a bipolar transistor, which will be described in the next section.

IV. Active Elements

In an integrated circuit, all connections are made at the top surface of an IC wafer, as opposed to discrete transistors. The transistors are isolated from other elements in the chip with a thermal oxide coating surrounding the transistor. The majority of bipolar transistors are of the npn type because the electrons have a higher mobility as minority carriers in the base region than do holes in a pnp configuration. This higher mobility is desirable since modern devices continually seek higher-speed operation.

To give the reader an idea of the manufacturing process involved in constructing a bipolar transistor, we will outline some of the major processing steps. A typical scenario for the fabrication of an npn transistor will be outlined. The starting material is a lightly doped p-type polished silicon wafer. The first step is to "bury" an n^+ layer into this wafer. This layer will reduce the series resistance of the collector. A thick oxide (≈ 1 μm) is thermally grown on this wafer, and a window is then opened in the oxide. Low-energy arsenic ions are implanted into this window. See Figure 9.3(a). The dimensions in these figures are not to scale.

The second step is to deposit an n-type layer. The oxide is removed and an n-type layer is grown in its place, as shown in Figure 9.3(b). The technique of growing a single oriented crystal on a substrate is called epitaxial growth. The thickness and doping concentration determine the ultimate use of the device. Digital circuits with their lower voltages require thinner layers than analog circuits and also have a higher doping level. Note in the figure that there is an indication that back-diffusion has carried some of this deposition back into the

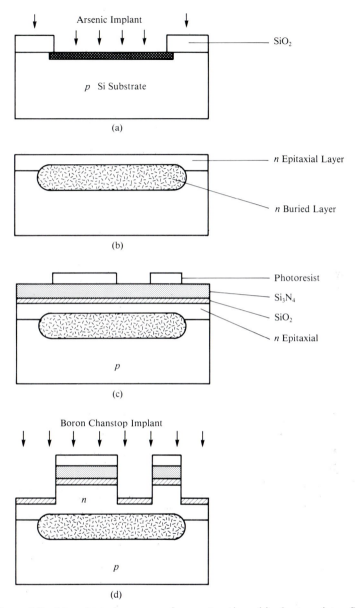

Figure 9.3   Manufacturing process for constructing a bipolar transistor. See text for explanation.

epitaxial layer. This can be minimized by using a low-temperature epitaxial process.

The third step is to thermally grow a lateral oxide isolation region followed by a silicon nitride deposition. The just deposited thin oxide layer will protect the silicon in later high-temperature processing steps. The nitride–oxide layers and a

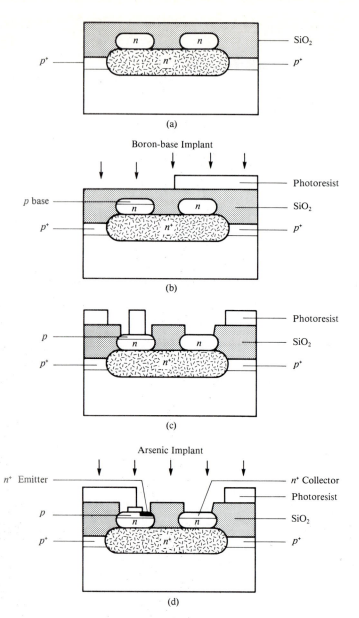

Figure 9.4   Continuation of Figure 9.3.

portion of the *n* layer are etched using a photoresist as a mark, as shown in Figure 9.3(c). A photoresist is a photosensitive organic material. Finally, boron ions are implanted into the exposed silicon areas, as noted in Figure 9.3(d).

The photoresist is now removed and the wafer is placed in a furnace, which will cause an oxidation to take place. Thick oxide layers will be grown in layers not covered by the nitride layers since the nitride layer has a low oxidation rate. This is shown in Figure 9.4(a). One added feature is that the implanted boron ions will be "segregated" underneath the isolation oxide to form a p^+ layer. This layer helps prevent surface inversion and eliminates high-conductivity paths among neighboring buried layers (i.e., a channel stop or "chanstop").

The fourth step is to form the base region. The right half of the device is protected with a photoresist mask, and boron ions are implanted to form the base region, as shown in Figure 9.4(b). A lithographic process removes most of the thin oxide pad except in a small region in the center of the base region. See Figure 9.4(c).

The fifth step is to form the emitter region. The base region is still protected with a photoresist. An arsenic dose is implanted to form the n^+ emitter and n^+ collector contact regions, as noted in Figure 9.4(d). The photoresist is now removed and the contacts to the base, emitter, and collector are finally metallized into place. This is summarized in the perspective drawing shown in Figure 9.5.

A bipolar transistor has now been fabricated, and it involves several separate steps of film-formation operations, lithographic operations, ion implantations, and etching operations. Failure at any stage of the process will cause the failure of the total operation.

In addition to fabricating a bipolar transistor in the preceding procedure, we have also fabricated two pn diodes. The bipolar transistor is, however, the dominant active element in the VLSI design area. The dominant device is the MOSFET, which will be discussed in the next section.

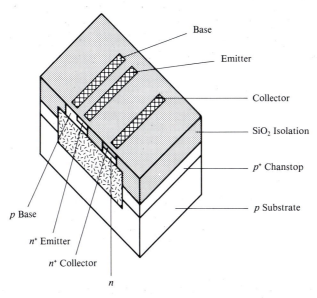

Figure 9.5 Perspective view of an oxide-isolated bipolar transistor.

V. MOSFET Fabrication

Since the MOSFET can be scaled to smaller dimensions than other types of devices, it has assumed a dominant position in the VLSI circuit. It has two distinct relatives, the NMOS which is an *n*-channel MOSFET and the CMOS. The CMOS provides both an *n* channel and a *p* channel on the same chip. As will be described later, the NMOS requires fewer processing steps than the bipolar transistor. The CMOS has a lower power consumption than either the NMOS or the bipolar transistor. It is common now to find a MOSFET with a cross-

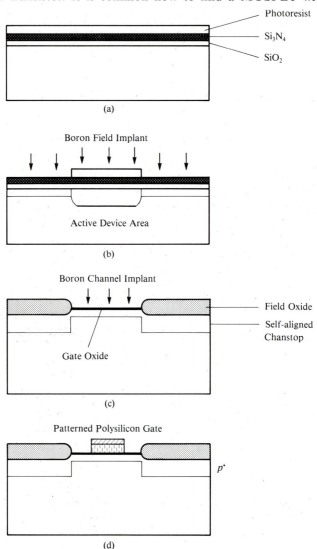

Figure 9.6 Manufacturing process for an NMOS. See text for details.

sectional area on the order of 50 μm^2, which enhances its application in VLSI circuit applications.

   The fabrication of an NMOS starts with a lightly doped p-type polished silicon wafer. The first step is to thermally grow a thin oxide layer. This is followed by a silicon nitride deposition, as shown in Figure 9.6(a). The desired active region

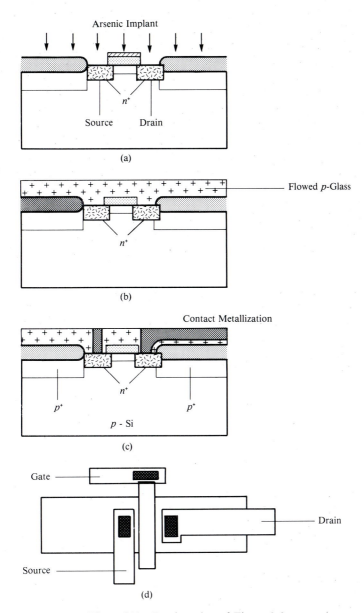

Figure 9.7   Continuation of Figure 9.6.

is defined by a photoresist mask. A boron layer is then implanted through the composite oxide–nitride layer. See Figure 9.6(b). The nitride layer, except for that portion covered by the photoresist mask, is then removed by etching. After the removal of the mask, the wafer is placed in an oxidation furnace to grow an oxide layer (called the field oxide) in the regions where the nitride is removed and to drive in a boron implant.

Next the gate oxide is grown and the threshold potential ϕ_T is set. The oxide–nitride layer over the active device area is removed and a thin oxide layer is grown. For a depletion-mode n-channel device, arsenic ions are implanted in the channel region and this decreases ϕ_T. Boron ions implanted in the channel region will create an enhancement mode and increase ϕ_T. See Figure 9.6(c).

The gate as shown in Figure 9.6(d) is formed by depositing a polysilicon layer, which is then heavily doped by diffusion or phosphorus implantation. For reduced resistance gates and smaller devices, a refractory metal such as molybdenum is frequently used.

The next series of operations is to form the source and the drain. A mask [Figure 9.6(d)] overlays the wafer and arsenic is implanted. There may be some parasitic capacitance connections between terminals, but this can be minimized if the operations are kept at a lower temperature to prevent lateral diffusion. This implantation is indicated in Figure 9.7(a).

The final operation is the metallization of the electrodes. A phosphorus-doped oxide is deposited over the entire wafer. By applying heat, this can be made to flow and cover the entire surface [Figure 9.7(b)]. Holes are etched in the surface and a metal such as aluminum is deposited, as noted in Figure 9.7(c). A top view of a MOSFET is shown in Figure 9.7(d).

As in the fabrication of the bipolar transistor, several stages are involved. If there is one incorrect operation at any stage, all is usually lost. The NMOS fabrication has three fewer opportunities than the bipolar transistor for such a possibility.

VI. Charge-Coupled Devices

In the devices described so far, an effort has been made to minimize any coupling between adjacent elements. It was assumed that any coupling would detract from the final performance characteristics of the device. In the present section, we will examine the operation of a device that relies on the coupling of charge from one region to another. The device is called a charge-coupled device and is given the acronym CCD. It has application as an analog or a digital delay line, a random-access memory for digital circuits, an analog signal processor, and an optical or thermal image sensor.

Let us examine the operation from the most elementary point of view, that of a series of MOS capacitors that are in juxtaposition. In fact, the metal electrodes are separate and isolated from each other, but the oxide and semiconductor are common to all electrodes, as shown in Figure 9.8. It is

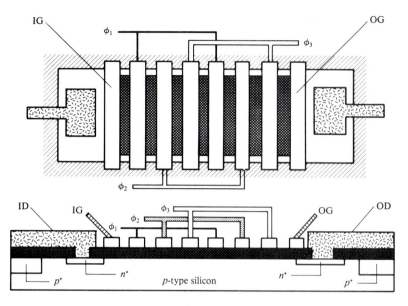

Figure 9.8   Charge-coupled device.

convenient again to choose a p-type semiconductor for analysis. Although isolated, a reverse-biased potential applied to one electrode will create a depletion layer under that electrode and a portion of a depletion layer under an adjacent electrode due to coupling.

The electrodes are connected to an external clock sequence of signal voltages ϕ_1, ϕ_2, and ϕ_3 in units of three electrodes per "cell." The next cell of three electrodes will be connected in the same sequence, so the cell will be the fundamental building block for a CCD. The value of the reverse-biased potential $\phi_j (j = 1, 2,$ or 3) determines the depth of the depletion layer under that particular electrode. In addition to the cells, an input diode (ID) and an input gate (IG) that will be used to inject charge into the cells and an output diode (OD) and output gate (OG) that detects the presence of a charge packet in the array of cells that comprise the main body of the CCD are also incorporated in the structure. The region of p^+ at either end prevents the inversion of the p-type silicon substrate. The main body of the CCD and additional input and output structures make the device shown in Figure 9.8 appear to be a multigate MOS transistor.

From the clock sequence given in Figure 9.9(a), we note that at a time t_b, the electrodes labeled 1 are turned on ($\phi_1 > \phi_T$) and an inversion layer exists under that particular electrode. The voltages applied to electrodes 2 and 3 are equal to zero or at least less than ϕ_T, so they are not in the inversion mode (off). Both the input diode and the output diode shown in Figure 9.8 are biased with high positive voltages to prevent the inversion of the p regions under the gates. Hence the surfaces under both gates are in deep depletion and neither diode can supply electrons into the main CCD array. As a result, the MOS capacitors that comprise the main CCD array are also in deep depletion, and the surface

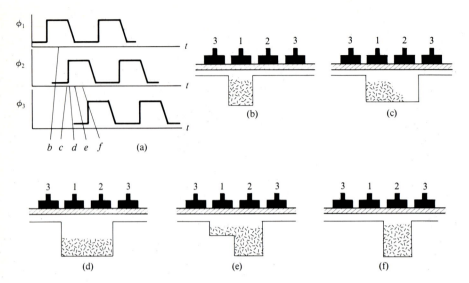

Figure 9.9 Charge-coupled device. (a) Timing sequence of potentials applied to the three-electrode cell. (b–f) Motion of charge packet under the electrodes. See text for detailed sequence of operation.

potential will be determined by these voltages applied to the electrodes. In Figure 9.9(b), the surface potential under electrode 1 will be higher than under electrodes 2 and 3.

Since the distance between adjacent electrodes is sufficiently small, the transition of surface potentials under one electrode to the adjacent electrode will be smooth even if the electrode is higher than the adjacent one. This will create energy wells under electrodes labeled 1. If the CCD remains in this state for a long time, then thermally generated electrons will fill up the energy well, which will form a charge packet that is confined to the region under electrode 1 at a time t_b.

However, we normally operate with a clock sequence that is sufficiently fast that the contribution of the thermally generated electrons is small compared with those injected into the CCD. At the time of injection, the voltage of the input diode is lowered to be between the surface potentials under the input gate and the ϕ_2 electrode. Electrons will see a region of lower potential and will flow into the region under the ϕ_1 electrode through the input gate. The injection of these electrons will be stored under the input gate and the first ϕ_1 electrode, and the surface potentials under the input gate and this electrode will be the same as the input diode voltage. At a later time, the input diode is shut off by returning its voltage to a high value. Any excess electrons will be extracted through the input diode lead. Hence a well-defined charge packet has been created under the first ϕ_1 electrode. Its size is proportional to the difference of surface potential under the input gate and that of this electrode.

At a time t_c, the potential is applied to both electrodes 1 and 2, which creates a depletion layer under both electrodes, and the charge will diffuse under both electrodes as shown in Figure 9.9(c). At a time t_d, the charge will be uniformly distributed under both electrodes since the applied potential is still applied to both electrodes. The potential applied to electrode 1 is now decreased such that at time t_e the depletion width decrease causes the stored charge to migrate to the region under electrode 2. Finally, at time t_f the voltage has returned to its unbiased condition and the charge packet now resides under electrode 2. It will remain there until it feels the effects of the potential applied to electrode 3, when the sequence starts over and repeats itself.

If the voltage switching of the clock sequence is slow such that a charge remains under two electrodes at the same time, this is sometimes called a "fat zero" if the charge presence is used to indicate a zero. We usually also examine the opposite temporal operation on how fast a CCD can operate. The fat zero will reduce the dynamic range of the CCD. This is determined by the speed with which we can transfer the charge from under one electrode to the next.

VII. Conclusion

The procedure of fabricating integrated circuits involves close cooperation between the engineer who designs the electronic circuits, the physicist who has developed the semiconducting materials, the chemist who produces these materials, and the marketing person who sells them. In devices whose size is approaching ultimate limits, the amalgamation of science and technology is reaching its modern-day zenith. This chapter has just brushed the technological revolution that has occurred and is occurring in the last third of the twentieth century. The reader has no choice at this time but to keep studying since changes are happening almost daily. For this chapter, which is very qualitative, no problems are given. The reader is now encouraged to spend an equivalent time in the library and seek out some new results and techniques in journals. It will be time well spent.

Chapter Ten

Additional Solid-State

Effects and Devices

The cupboards containing new solid-state effects and their eventual metamorphosis into practical devices are not yet bare, and the reader will find much from which to still choose even after the diode, transistor, and integrated-circuit crowd have satiated their enormous appetites. Rather than just feed the masses, there are certain gourmet items that remain to be savored before leaving the smorgasbord table that was first set by the solid-state chefs after they walked in the footsteps of the giants who had first stoked the ovens. These items all have their own flavor, and the reader will be able to obtain a sample here as the dishes are passed by. It is hoped that a taste will cause the true gourmet to study a particular delicacy with more fervor than can be obtained from this short and limited menu of haute cuisine.

Several devices are based on an application of the negative conductance that was noted first in the study of a tunnel (or Esaki) diode. These devices are not based on the transport of charge over a finite distance, as was noted in the operation of the other solid-state devices. This will allow the devices to be operated at higher frequencies. These devices have gradually replaced many cumbersome tubes in low-power microwave applications. Hence the "computer on the wrist" could be connected to the "radar in the hand" by the local gendarmes as they enforce local speeding mandates. Other devices that could possibly have certain computer applications would lead to developments among the mathematical aficionados who would try and explain various observations in these new devices. The word **soliton** would arise during an explanation of certain of these devices and an entire new restaurant would be opened for further study to satisfy a new set of diners.

I. IMPATT Diode

The application of either a tunnel diode, a varactor diode, or a specially designed transistor in a microwave circuit is limited to a frequency range that is determined by the transit time of the charged particles T. This is typically on the order of $1/T \approx 10^9$ hertz (Hz). Other physical mechanisms will allow for the generation of high-frequency signals, and one of the most important is the application of instabilities that may exist in semiconductors. One of these instabilities involves the *negative conductance* that can be found under certain conditions in semiconductor applications. Two devices that make use of this instability are the *IMPATT* diode and the *Gunn* diode. Both devices can be used at microwave frequencies to provide amplification of a signal or a high-frequency oscillation that can be used as a source of energy. The first of these devices will be described in this section.

The IMPATT diode incorporates the physical effects of impact avalanche ionization and transit time effects (the name IMPATT follows from IMPact Avalanche and Transit Time). Simple *pn* diodes will exhibit an avalanche breakdown if they are reverse biased. If a small ac signal is superimposed on the reverse-bias dc voltage, the ac perturbation of charge carriers will be swept through a drift region by the dc avalanche. The ac component of the resulting current will be almost 180° out of phase with the applied voltage, and this results in a negative conductance. IMPATT diodes can convert dc energy into energy at this high frequency very efficiently, in particular, if connected in a well-designed external circuit.

The first suggestion of using an IMPATT configuration as a microwave device was made by W. T. Read in 1958. It was demonstrated seven years later. There are several possible configurations for the semiconductors other than the one proposed originally by Read, which was $n^+ pip^+$, where *i* is an intrinsic region. Another diode is the $p^+ nn^+$ (or *pin*; heavily doped *p* and *n* regions sandwiching a lightly doped *n* region) silicon IMPATT diode, which will be analyzed here since it is one of the most commonly used diodes. It is shown in Figure 10.1(a). The device is biased with a dc value ϕ_A that is just less than the critical value, which will cause breakdown. The *n* region is doped such that the *n* region is fully depleted or "punched through" at breakdown. Also, this value of doping is such that the electric field at breakdown will cause the velocity to saturate at a value v_d. The transit time *T* for the electrons to traverse the *n* region is defined by

$$\tau = \frac{L}{v_d} \tag{1}$$

where v_d is the saturated value of the drift velocity and *L* is the width of the *n* region. Let us assume that a very small ac voltage of frequency ω_0 (where $\omega_0 T = \pi$) is added to the dc bias voltage ϕ_A. See Figure 10.1(b). The resulting electric field that appears in the *n* region will be modulated such that at certain times it will have a value that is above the critical value required to cause the diode to avalanche, as noted in Figure 10.1(e). Let us call this time t_0 as being the time when it just starts to avalanche.

The corresponding electron distributions at the various times of the period $2T$ of the ac signal are shown in Figures 10.1(f) through (h). The dashed lines represent the density from the previous cycle. For values of $t_0 + \Delta \tau \geq t \geq t_0$,

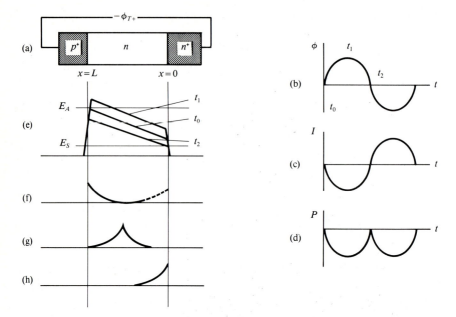

Figure 10.1 (a) IMPATT diode. (b) Applied ac voltage to the diode. (c) AC current. (d) AC power. (e) Electric field at various times of the applied ac voltage. (f–h) Excess electron density at various times.

where $\Delta\tau$ is the time interval that the voltage is greater than ϕ_A, the diode will experience avalanche multiplication of the reverse saturation current, which will begin at the high field region at $x = L$ as noted in Figure 10.1(e). An impact ionization may occur, which will create a hole–electron pair at $x \approx L$. The applied field causes the hole to drift into the p^+ region, where it recombines almost instantly. The electrons that are created in this ionization will drift to the region $x < L$. They will drift with a velocity $\approx v_d$. During the time interval $\Delta\tau$, the carrier density at $x = 0$ continues to grow exponentially, and the carriers generated through the avalanching continue to drift toward $x = 0$. This is shown in Figure 10.1(f). At the time $t = t_0 + \Delta\tau$, the voltage drops below the critical value ϕ_A and the carrier density starts to decay to zero. The generated electrons continue to drift toward $x = 0$ and the current decays. During the time interval $t_0 + \Delta\tau \le t \le 2t_0 + \Delta\tau$, no new hole–electron pairs are created. Those that are already created continue to drift toward $x = 0$ as noted in Figure 10.1(g). The cycle repeats at $t = 2\tau$, as shown in Figure 10.1(h).

To understand the generated current, we start with the equation of continuity with the assumption of negligible recombination of electrons in the n region, which is written as

$$\frac{\partial j}{\partial x} + \frac{\partial (qn)}{\partial t} = 0 \tag{2}$$

This can be integrated over the drift region:

$$j(L) - j(0) = - \int_0^L \left[\frac{\partial(qn)}{\partial t} \right] dx \tag{3}$$

The derivative that appears within the integral can be approximated with the relation

$$\frac{\partial(qn)}{\partial t} \approx \frac{qn}{\tau}$$

Therefore, (3) can be written as

$$I(t) \approx \frac{A}{\tau} \int_0^L qn(x, t) \, dx \tag{4}$$

where A is the cross section of the device. The maximum value for the current $I(t)$ will be found at a time when this integral is a maximum, which will occur at a time that is defined as $t = t_2$. The current is shown in Figure 10.1(c). The ac power, which is a product of the ac current and the ac voltage, which are shown in Figure 10.1(d), is *negative*. This implies that, once started, the oscillations will continue since energy is being absorbed by the device. Random fluctuations due to thermal noise can of course start this oscillator.

Analytical theories are available to model the behavior of an IMPATT oscillator. Although not entirely correct, the models may yield some valid and useful predictions. The descriptions usually start with three equations. The first is Poisson's equation

$$\frac{\partial E}{\partial x} = \frac{q}{\varepsilon}(N_d - N_a + p - n) \tag{5}$$

The other two are the separately written equations of continuity for holes,

$$q\frac{\partial p}{\partial t} + \frac{\partial j_p}{\partial x} = \alpha_n j_n + \alpha_p j_p \tag{6}$$

and electrons,

$$q\frac{\partial n}{\partial t} - \frac{\partial j_n}{\partial x} = \alpha_n j_n + \alpha_p j_p \tag{7}$$

In writing (6) and (7), the coefficients α_n and α_p, which are functions of the local electric field, refer to the average number of ionizing events per unit length. In addition, recombination has been neglected. The currents are related to the densities by

$$j_n = -qnv_d \quad \text{and} \quad j_p = qpv_d \tag{8}$$

where the electron and hole saturated drift velocities are assumed to be equal. There are three independent quantities, j_n, j_p, and E, in equations (5) to (7). The procedure that is followed is to make a small-signal approximation and write

$$E = E_0 + e_1 e^{i\omega t} \tag{9}$$

$$j_n = J_{n0} + j_{n1} e^{i\omega t} \tag{10}$$

$$j_p = J_{p0} + j_{p1} e^{i\omega t} \tag{11}$$

The resulting ratio

$$Y = A \frac{j_1}{\phi_1} \tag{12}$$

where

$$j_1 = j_{n1} + j_{p1} + i\omega e_1 = \text{constant}$$

and ϕ_1 is the voltage that created j_1, yields the admittance of the diode. The susceptance and the conductance are frequency dependent due to (1). The former changes signs and the latter has a negative sign; hence it is a source of energy. Several numerical steps are involved, so the details are not presented here.

To build an oscillator, the external circuit must be designed to provide a conjugate match for the admittance of the diode. Although in principle this will work, it usually does not yield the optimum performance. The derivation was made assuming small-signal operation, while the real-world situation has large-signal conditions. To describe this case, numerical solutions of the full set of nonlinear partial differential equations, which includes space-charge effects, are required. It may be possible, however, to suggest some analytical expectations using the procedure outlined in Appendix A. An example will illustrate this.

Example 10.1   Simplify the description of an IMPATT diode such that only the electron motion with no ionization will be considered and the holes can be neglected. Also, let the donor and acceptor densities be almost equal. Describe the transient motion of the electrons.

Answer: In this limit, Poisson's equation becomes

$$\frac{\partial E}{\partial x} = -\frac{qn}{\varepsilon}$$

and the continuity equation of electrons is written as

$$q\frac{\partial n}{\partial t} - \frac{\partial j_n}{\partial x} = 0$$

where $j_n = -qn\mu_n E$ and μ_n is the mobility of the electrons. Eliminating the density between these three equations, we obtain the nonlinear partial differential equation

$$\frac{\partial(\partial E/\partial x)}{\partial t} + \mu_n \frac{\partial[E(\partial E/\partial x)]}{\partial x} = 0$$

Let the independent variables x and t be combined as $\xi = x/t$ (see Appendix A for the procedure to select this combination) and obtain

$$(\mu_n E - \xi)\frac{d^2 E}{d\xi^2} - \frac{dE}{d\xi} + \mu_n\left(\frac{dE}{d\xi}\right)^2 = 0$$

The first integral of this equation is

$$\mu_n E \frac{dE}{d\xi} - \xi\frac{dE}{d\xi} = k_1$$

Setting the constant of integration $k_1 = 0$, we find the second integral to be

$$E = \frac{\xi}{\mu_n} = \frac{x}{t\mu_n}$$

Therefore, the density and current densities are computed from

$$n = -\frac{\varepsilon}{q\mu_n t} \quad \text{and} \quad j_n = \frac{\varepsilon x}{\mu_n t^2}$$

This "space-charge limited current" problem has also received attention among scientists examining the charge motion in insulating fluids.

The key parameters that determine the operation of an IMPATT diode are the distribution of the electric field, the ionization coefficients, the saturated drift velocity, and the carrier lifetime, which must be greater than the particle transit time T. If the breakdown voltage increases with increasing temperature, heat will be distributed evenly across the diode and hot spots, which might cause filament formation, will not develop.

II. TRAPATT Diode

The TRAPATT oscillator, which is an acronym for TRAPped, Plasma, Avalanche, Triggered Transit, was experimentally demonstrated prior to its modeling in 1969. The power output of such a device can be in the range of hundreds of watts at a few gigahertz. The IMPATT device described previously operates with powers of watts at frequencies up to 50 GHz. The physical understanding of the operation of a TRAPATT oscillator follows from the discussion of the IMPATT device, although there are significant differences. The major theoretical difference is that an analysis of an IMPATT oscillator follows from a small-signal model in that we can perturb about some equilibrium value to obtain the characteristics. The TRAPATT operation only occurs in large-signal cases, so a similar perturbation scheme is not valid. It is found that TRAPATT devices operate at a lower frequency and with a high efficiency ($\approx 75\%$).

The basic physical process that occurs is similar to the IMPATT with the following major difference. If the voltage increases in an IMPATT such that avalanching takes place, the TRAPATT mode assumes that a large density of electrons and holes are *trapped* and move together as a packet. This "plasma" suppresses the electric field, and their velocity will be less than the saturated value for an IMPATT diode. Therefore, the frequency of operation will be lower.

It is possible to present an analysis of the operation of a TRAPATT by examining an n^+np^+ abrupt junction device as shown in Figure 10.2(a). The electric field at $t = 0$ will have a value to just punch through, but will be well below the value required to cause avalanching. The electric field can be computed from Poisson's equation

$$\frac{\partial E}{\partial x} = \frac{qN_d}{\varepsilon} \tag{13}$$

This is valid under the conditions of the electric field being beneath the value to cause avalanching since there are no mobile charges present. If the electric field at $x = 0$ is taken to be zero, the integral of (13) yields

$$E(x) = \frac{qN_dx}{\varepsilon} \tag{14}$$

which is shown in Figure 10.2(b).

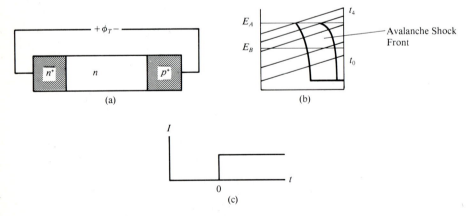

Figure 10.2 (a) TRAPATT diode. (b) Electric field across the diode. When this field is greater than the critical value, avalanching occurs and a shock front propagates. (c) Current output of the diode.

Let us now connect a current generator to the diode that is turned on at $t = 0$ to have a current density j_0 as shown in Figure 10.2(c). As long as the electric field remains below the threshold value required for avalanching ($E < E_A$), no mobile charges are present in the diode and the only current that can flow in the diode is a displacement current, which is given by

$$j_0 = \varepsilon \frac{\partial E}{\partial t} \tag{15}$$

The integral of (15) is

$$E(x, t) = E(x, 0) + \frac{j_0 t}{\varepsilon} \tag{16}$$

This states that the electric field will increase everywhere in time. From (14), we know the value of the electric field at $t = 0$:

$$E(x, 0) = E(x) = \frac{q N_a x}{\varepsilon}$$

Therefore,

$$E(x, t) = \frac{q N_a x}{\varepsilon} + \frac{j_0 t}{\varepsilon} \tag{17}$$

A point of constant electric field, say E_B in Figure 10.2(b), will move to the left with a velocity given by

$$\Delta E_B = 0 = \frac{q N_d}{\varepsilon} \Delta x + \frac{j_0}{\varepsilon} \Delta t$$

or

$$v_x = \frac{\Delta x}{\Delta t} = -\frac{j_0}{q N_d} \tag{18}$$

If j_0 is large enough and N_d is small enough, this velocity may exceed the saturated drift velocity.

As time increases, we note from (17) that the electric field will increase in value until it exceeds the critical value E_A required to cause the diode to avalanche. For $E > E_A$, (17) is no longer valid since the field can no longer increase due to the sudden increase in the local carrier density due to the avalanche multiplication. With v_x being much larger than the saturated value for the drift velocity for electrons and holes, the particles do not have sufficient time to equilibrate. Hence the electric field can become much greater than E_A. However, the avalanche-induced carriers suppress the electric field to a very low value behind the peak as it

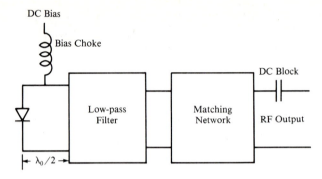

Figure 10.3   Equivalent circuit of a TRAPATT diode.

is swept to the left in Figure 10.2(b). This front that moves from right to left is called the *avalanche shock front*. Typical transit times L/v_x are on the order of tens of picoseconds.

The terminal voltage that equals the spatial integral of the electric field changes from a high value ϕ_1 above the avalanche breakdown voltage to a very low value $\phi_2 \approx 0$ V in a time equal to this transit time. If the diode is connected to a transmission line whose characteristic impedance is Z_C, as shown in Figure 10.3, a current pulse I, which changes from $I \approx 0$ to $I \approx \phi_1/Z_C$, will be excited. In this process of switching, the "trapped" charges will be extracted by this current and it will decay back to the lower value. It will be recharged through an external dc bias and the cycle will be repeated. In Figure 10.3, the diode is located a distance $\lambda_0/2$ from a low-pass filter, which is tuned to pass only the fundamental frequency component $f_0 \approx (L/v_x)^{-1}$. Once started, the TRAPATT mode efficiently converts dc bias energy into high-frequency energy almost indefinitely.

Although the IMPATT mode is a critical requirement for the starting of the TRAPATT mode, it quickly loses its identity and the two modes of operation are significantly different in character. The frequency of operation of an IMPATT is several times that of a TRAPATT, and the TRAPATT mode has sometimes been described as a subharmonic IMPATT diode, although the modes are entirely different.

III. Gunn Diode

The basic idea of a Gunn diode (or *transferred-electron* mechanism) is to shift electrons that are in the conduction band from a state of high mobility to a state of low mobility using a strong electric field. If this mechanism occurs, then a region of negative conductance operation can be found. To understand the mechanism, recall the discussion of energy bands in Chapter 4. It is reasonable to examine the energy bands as a function of configuration space when the conduction electrons reside near the minimum energy of the conduction band. However, in more realistic situations, the electron energy should be plotted as a

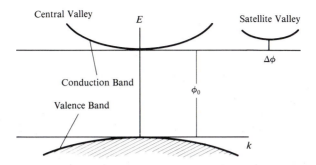

Figure 10.4   Semiconductor energy-band diagram.

function of the momentum. Since the momentum is proportional to the wave vector **k**, it is therefore prudent to examine the energy as a function of **k**.

In Figure 10.4, a simplified band diagram for a semiconductor is shown. In this case, we have modeled a GaAs *n*-type semiconductor, where the valence band is normally filled and the central valley, which is the minimum of the conduction band, is at **k** = 0. This band normally contains conduction electrons. In addition, additional valleys exist called satellite valleys, which occur at a higher energy and at a location in momentum space (or **k** space) that is different from zero. These satellite valleys differ by several $k_B T$ from the central valley and are normally unoccupied. Under the influence of a very large externally imposed electric field, which is greater than some critical value E_C, the electrons that are resident in the central valley may gain sufficient energy to move into one of these satellite valleys.

As long as this external electric field is imposed on the material and its value is greater than the critical value, the electrons will remain in the satellite valley. The effective density of states in this satellite valley is greater than the central valley. Hence the electrons, which will be scattered among different states, will most probably remain in the region with the most states rather than scatter back to the central valley, which has fewer states as long as $E > E_C$. The effective mass of these electrons will then be determined by the satellite valley rather than the central valley. It may have a value there that is considerably different from its value when $E < E_C$. For example, the effective mass of electrons in the satellite valley is almost 20 times its value in the central valley. Therefore, its mobility μ will be much lower since $\mu \approx e\tau_C/m^*$.

Example 10.2   Calculate the ratio of the effective density of states in the upper and lower levels of a Gunn diode.

Answer: The effective density of states N_C is defined in Chapter 5 by the relation

$$N_C = 2\left(\frac{2\pi m_n^* k_B T}{h^2}\right)^{3/2}$$

Therefore, the ratio of states is given by

$$\frac{N_U}{N_L} = \left(\frac{m_n^*(U)}{m_n^*(L)}\right)^{3/2}$$

Since the effective masses differ by approximately 20, this ratio becomes $(20)^{3/2} \approx 89$.

This effect has an important consequence for the electron motion under the influence of an electric field. For values of $E < E_C$, the velocity of the electron will increase as the field increases. However, at the critical value where $E = E_C$, the electrons will start to *slow down* as the electric field is further increased. Therefore, these electrons can gain potential energy at the expense of kinetic energy since the velocity is decreasing at these particular values of electric field. The current density j is defined from

$$j = nqv_d$$

so the differential conductivity σ, which is defined from

$$\sigma = \frac{dj}{dE}$$

may actually be negative.

A possible relation between the electron velocity and the electric field is shown in Figure 10.5. The asymptotic values of the mobilities when the electrons are in the central valley (μ_L) and the satellite valley (μ_H) are also given. Between these two states, a region of negative differential mobility exists.

This region of negative mobility will have some important consequences. First, the conductivity $\sigma = nq\mu$ will be negative. Second and more important, space charge instabilities will occur and the device cannot be maintained in a dc stable state in this region of electric field space. This can be shown from the

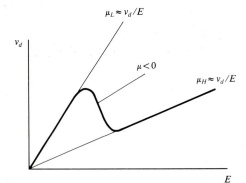

Figure 10.5   Electron velocity versus electric field in a Gunn diode.

equation of continuity

$$\frac{\partial \rho}{\partial t} + \frac{\partial j}{\partial x} = 0 \tag{19}$$

Poisson's equation

$$\frac{\partial^2 \phi}{\partial x^2} = -\frac{\rho}{\varepsilon} \tag{20}$$

and Ohm's law

$$j = \sigma E = -\sigma \frac{\partial \phi}{\partial x} \tag{21}$$

From (20) and (21), we write

$$\frac{\partial j}{\partial x} = -\sigma \frac{\partial^2 \phi}{\partial x^2} = \frac{\sigma \rho}{\varepsilon} \tag{22}$$

Substituting this into the equation of continuity (19), we finally obtain

$$\frac{\partial \rho}{\partial t} = -\frac{\sigma \rho}{\varepsilon} \tag{23}$$

The solution of equation (23) is

$$\rho = \rho_0 e^{-t/\tau_d} \tag{24}$$

where $\tau_d = \varepsilon/\sigma$ is called the dielectric relaxation time. In normal dielectrics, random fluctuations of a charge density will quickly neutralize in a time given by τ_d, so an assumption of charge neutrality is a good one for most materials.

With the conductivity σ being negative, however, this implies that a charge perturbation can *grow* in time; in fact, a space-charge fluctuation will grow exponentially in time instead of decaying to zero. A dipole charge will be created at a localized region in space, as shown in Figure 10.6. Instead of dying away to zero, this dipole will *grow* in time due to the negative conductivity. The growth takes place in a region where the electrons are drifting due to the imposed electric field. Eventually, the drifting domain will reach the end, where it gives up its energy as a pulse of current to the external circuit. The frequency of operation f_{Gunn} of the Gunn diode will be determined by the drift velocity v_d and the length of the diode L. It can be written as

$$f_{\text{Gunn}} \approx \frac{v_d}{L}$$

which is the inverse of the transit time of the domain.

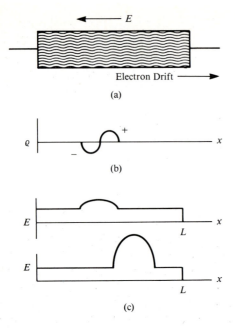

(a)

(b)

(c)

Figure 10.6 (a) Gunn diode with electron drift indicated. (b) Space-charge formation. (c) Growth of domain of electric field.

It is unlikely that more than one domain will form at a time, since a large fraction of the applied voltage will appear across this one domain at the expense of the electric field in the rest of the diode, which will decrease below the critical value required for negative conductivity. The field outside the domain will stabilize at a value of positive conductivity (point A in Figure 10.7) and the field within the domain will stabilize at a value of negative conductivity (point B in Figure 10.7).

To explain the formation of a single domain, we will assume that a small dipole has formed in the diode, possibly due to a local crystal defect, doping inhomogeneity, or a local noise fluctuation. This dipole grows and drifts down the bar as a domain. Except for a small increased perturbation in the dipole layer, we can assume that the field is approximately uniform at point C in Figure 10.7, which is in the negative mobility region. Within the dipole, the electron drift velocity will be slightly less than outside of it. Hence electrons will pile up on the upstream side and will be depleted on the downstream side. This further increases

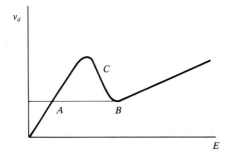

Figure 10.7 Velocity-field characteristics for an *n*-type Gunn diode. Points A and B are stable points. Point C is unstable.

the dipole field within the domain, and the field outside the domain decreases. A stable point is reached when the field within the domain is at point B and the field outside is at point A. At these values, the domain will drift down the rod at a constant velocity v_d.

In the preceding discussion, we have assumed that the rod was long enough that the domains had sufficient "room" to grow to the saturated value. This implies that the transit time L/v_d must be greater than the absolute value of the dielectric relaxation time ε/σ. Therefore,

$$\frac{L}{v_d} > \frac{\varepsilon}{\sigma} = \frac{\varepsilon}{nq\mu} \tag{25}$$

The first observation of this effect was by J. B. Gunn, who measured the current voltage characteristics of a GaAs sample. Up to a critical value of bias voltage, a linear ohmic relation was detected. Above this critical value, current pulses were detected. The pulses were separated in time by a transit time L/v_d. In addition, using a small capacitive probe to monitor the local electric field along the rod, Gunn was able to demonstrate the formation and growth of the localized domain as it drifted with a velocity v_d.

In addition to the Gunn effect, there are several other modes of operation of a diode based on the basic premise of transferred-electron devices. It is also not the most desirable mode of operation for many applications. The resulting current pulses are inefficient sources of microwave power. One possible mode that has a higher efficiency is the *limited-space-charge accumulation* (LSA) mode. In this mode, domains do not have enough space or time to form in the transferred-electron device. The diode is connected to a high-frequency tuned circuit, the tuning frequency being much higher than (transit time)$^{-1}$. Since the field is above the threshold value, the diode is maintained in the negative conductance state for a large fraction of a cycle. During the small time that the diode is below threshold, the accumulation of electrons should have sufficient time to collapse.

IV. Josephson Effects

The Josephson effect has been proposed as an effect that could have many practical applications in devices where information is either to be stored or processed. It is based on an understanding of the quantum nature of superconductivity. Certain characteristics of the effect are presented here. We find that it is also possible to manipulate the equations that describe the effect and turn them into an equation that is one of a set of *nonlinear evolution equations* that have a **soliton** as a solution. The soliton is an entity that has and is continuing to receive considerable attention among a certain community of investigators. One property that this community has uncovered is that nonlinear solitons are very stable objects and do not destroy themselves upon interacting or colliding. Hence we can speculate about storing information on one of them and it would not be forgotten even if it collided with another one in a computer memory. We will describe the effect and show the proper manipulation required to obtain the soliton-bearing equation in Chapter 11.

The Josephson effect is an important departure from the effects that have been studied up to this point. The *pn* diode and the transistors that were described previously could be modeled with a classical model. The Josephson effect requires a quantum description and is observed in certain superconductors. The junction capacitance that arises from the classical model will also be important in describing its operation.

The discovery of this phenomenon could also serve as a stimulus to students (and teachers) since in his lecture at the time of receiving the Nobel prize Brian Josephson recalled the stimulating atmosphere at the laboratory during his research student days in the early 1960s when he predicted the effect.

A Josephson junction is shown in Figure 10.8. Two superconductors are separated by a narrow insulating barrier, which is typically an oxide with a thickness on the order of 10 Å. The thickness of the *superconductors* may be several thousands of angstroms thick. Normal vacuum deposition techniques are employed to construct this "sandwich." The theory of superconductivity is due to Bardeen, Cooper, and Schrieffer (BCS) and employs ideas from quantum mechanics. A brief summary will be given next so that the reader can gain some insight into the operation of the Josephson effect. This effect is a quantum effect and not a classical effect.

The starting point of the calculation is to examine the wave functions and some of the definitions described in Chapter 1. In particular, the electric current can be defined in one dimension in terms of the wave function ψ as

$$j = \frac{q}{2m}[\psi(p\psi)^* + \psi^*(p\psi)] \tag{26}$$

where $p = -i\hbar\partial/\partial x$ is the momentum operator, q is the charge, and the notation * represents the complex conjugate of the function. The wave function ψ is a solution of the Schrödinger equation. The particle density $\rho(x)$ at a location x is defined as

$$\rho(x) = \psi\psi^* \tag{27}$$

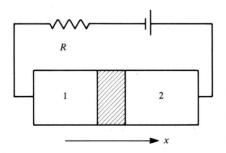

Figure 10.8 Schematic of Josephson junction consisting of two superconductors separated by an insulating barrier.

The wave function ψ is a complex function that can be written in phasor notation as

$$\psi = \sqrt{\rho(x)}\, e^{i\theta(x)} \tag{28}$$

where both $\rho(x)$ and $\theta(x)$ are real functions. If this is substituted in (26) and the momentum operator is employed, we obtain

$$j = \frac{q}{2m}\left\{\sqrt{\rho(x)}\, e^{i\theta(x)}(i\hbar)\frac{\partial[\sqrt{\rho(x)}\, e^{-i\theta(x)}]}{\partial x} + \sqrt{\rho(x)}\, e^{-i\theta(x)}(-i\hbar)\frac{\partial[\sqrt{\rho(x)}\, e^{i\theta(x)}]}{\partial x}\right\}$$

or

$$j = \frac{q}{m}\,\hbar\rho(x)\frac{\partial\theta(x)}{\partial x} \tag{29}$$

Equation (29) states that there will be a flow of current that is proportional to the change of the phase of the wave function. This is a critical observation for the understanding of the Josephson effect. Since this current is related to the momentum of the particle by

$$j = \rho q v = \frac{\rho q}{m}\,mv = \frac{\rho q}{m}\,p$$

then the momentum p of the particle will also be proportional to the change of the phase of the wave function.

An analogy could be made at this point to the pastry chef who is trying to decorate a cake using a cloth bag containing the icing, which is connected to the spout. By twisting the bag, the icing is forced out through the small hole in the spout and onto the cake.

In a normal metal, noninteracting electrons would fill up any available states up to an energy equal to the Fermi energy E_F. One feature of superconductivity that is unique is that there is an *attractive interaction* between pairs of electrons as the temperature $T \to 0$. These pairs are called Cooper pairs, and the participating electrons occupy the states with opposite wave vectors and spin. This Cooper pair will behave as a single particle and we will treat it as one in the following discussion.

Let us consider a superconductor to which no external magnetic field is applied and in which there is no center-of-mass motion of the Cooper pairs (i.e., there is no directed drift). This implies that there is no current and, from (29), we write with $j = 0$ that

$$\frac{\partial\theta(x)}{\partial x} = 0 \tag{30}$$

Therefore, the phase $\theta(x)$ is independent of position in a superconductor in which no current flows.

If a constant current flows, then (29) becomes

$$\frac{\partial \theta(x)}{\partial x} = C \tag{31}$$

which can be integrated to yield

$$\theta(x) = Cx + \theta_0$$

where C and θ_0 are constants.

Since this attractive interaction of electrons exists in superconductivity, it is possible to just consider a superconductor as a collection of Cooper pairs that are all in the same state and have the same wave function, which is given by

$$\psi(x, t) = \sqrt{\rho(x, t)}\, e^{i\theta(x, t)} \tag{32}$$

The pair density is given by $\psi \psi^* = \rho$. The phase of the wave function is related to the current density of Cooper pairs in the superconductor via (29).

To now understand the operation of a Josephson junction, we will follow the procedure of first setting up the equations that describe the Cooper pair wave functions ψ in the two superconducting regions of Figure 10.8. In addition, we will assume that the pairs can "tunnel" through the insulator. Then we will consider a dc current passing through the junction and observe its effect on the phase of the resulting wave function.

The time-dependent Schrödinger wave equations that describe the wave functions in the two regions 1 and 2 in Figure 10.8 can be written as two coupled equations. This suggestion is similar to treatment of two separate electrical circuits that are coupled to each other with the "mutual coupling" found in, say, a transformer.

$$i\hbar \frac{\partial \psi_1}{\partial t} = U_1 \psi_1 + \hbar \Gamma_{21} \psi_2 \tag{33}$$

$$i\hbar \frac{\partial \psi_2}{\partial t} = U_2 \psi_2 + \hbar \Gamma_{12} \psi_1 \tag{34}$$

In this set of equations, U_1 and U_2 correspond to the lowest state energies of the superconductors, which we will take to be equal and to be much less than the contribution Γ_{12} and Γ_{21} that is due to the tunneling that occurs from one side to the other. As a simplification, this tunneling is taken to be isotropic in this discussion. Therefore, $\Gamma_{12} = \Gamma_{21} = \Gamma$. This essentially measures the transfer of Cooper pairs from one side to the other. The reader notes the analogy of this coupling of the wave function from one region to another to the problem of inserting a lossless dielectric slab across a waveguide. If an electromagnetic wave is incident from the left, a portion of it will be reflected from the slab, and the remainder will pass through the slab and appear on the right side. The electromagnetic wave, however, does not "tunnel."

With these approximations, the coupled pair of Schrödinger equations (33) and (34) simplify to

$$i\hbar \frac{\partial \psi_1}{\partial t} = \hbar \Gamma \psi_2 \qquad \qquad (35)$$

$$i\hbar \frac{\partial \psi_2}{\partial t} = \hbar \Gamma \psi_1 \qquad \qquad (36)$$

Recall that the existence of tunneling requires that a portion of the wave function from side 1 appears in side 2 also as shown in Figure 10.9. If there is no tunneling ($\Gamma = 0$), the wave functions computed from the now uncoupled equations (35) and (36) are independent of time, as was to be expected.

Let us define two complex wave functions of the form given in (32) to describe the Cooper pairs as

$$\psi_1 = \sqrt{\rho_1(x)} e^{i\theta_1(x)} \qquad \qquad (37)$$

$$\psi_2 = \sqrt{\rho_2(x)} e^{i\theta_2(x)} \qquad \qquad (38)$$

where the subscripts refer to the appropriate side of the Josephson junction and the dependence on time t is understood. These wave functions are now to be introduced into (35) and (36). Equating the real and the imaginary parts allows us to derive a set of equations for $\rho_{1 \text{ or } 2}$ and $\theta_{1 \text{ or } 2}$. The phases θ_1 and θ_2 will be connected to the current density j, which is given in (29).

Prior to actually substituting these functions into the set of coupled Schrödinger equations and performing the detailed calculation, let us try to predict the behavior of the expected results. This can be done with the simple electrical circuit shown in Figure 10.10, where a battery and resistor are connected to the Josephson junction. Let us assume that a current density j is created in the junction by the battery. Since $j \neq 0$, we know from (29) that

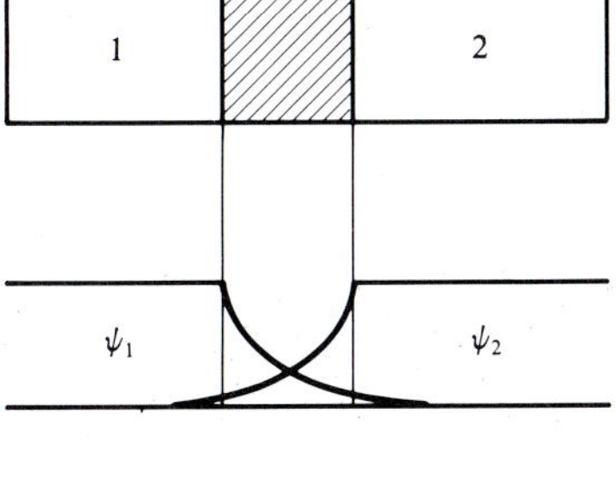

Figure 10.9 Schematic of a Josephson junction with the indicated wave functions.

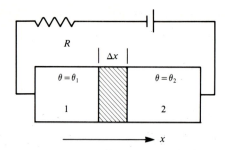

Figure 10.10 Current flowing through a Josephson junction with a phase gradient across the oxide barrier.

$d\theta/dx \neq 0$, and this current density has a value that can be computed from

$$\frac{d\theta}{dx} = \frac{mj}{\rho q\hbar} \tag{39}$$

The phase gradient will appear almost entirely across the oxide barrier. Note that the phase gradient is inversely proportional to the Cooper pair density ρ for a given value of current density j. This density is much larger in the superconductor than in the oxide by several orders of magnitude. Hence we can approximate the phase change as

$$\frac{d\theta}{dx} \approx \frac{\Delta\theta}{\Delta x} \approx \frac{\theta_2 - \theta_1}{\Delta x} \tag{40}$$

and this phase change appears across the oxide layer whose thickness is Δx. Therefore, a flow of current j through the junction produces a change of phase $\Delta\theta = \theta_2 - \theta_1$ across the oxide barrier layer.

The mathematical procedure to show this in detail is to substitute (37) and (38) in (35) and (36) and equate the real and imaginary parts. This procedure leads to four equations, which are written as

$$\frac{\partial\rho_1}{\partial t} = 2\Gamma\sqrt{\rho_1\rho_2}\,\sin\Delta\theta \tag{41}$$

$$\frac{\partial\rho_2}{\partial t} = -2\Gamma\sqrt{\rho_1\rho_2}\,\sin\Delta\theta \tag{42}$$

$$\frac{\partial\theta_1}{\partial t} = -\Gamma\sqrt{\frac{\rho_2}{\rho_1}}\cos\Delta\theta \tag{43}$$

$$\frac{\partial\theta_2}{\partial t} = -\Gamma\sqrt{\frac{\rho_1}{\rho_2}}\cos\Delta\theta \tag{44}$$

Let us now make the simplifying assumption that the superconductors on either side of the insulator are the same. This means that

$$\rho_1 = \rho_2 \tag{45}$$

From (43) and (44), we find that

$$\frac{\partial(\theta_2 - \theta_1)}{\partial t} = \frac{\partial \Delta\theta}{\partial t} = 0 \tag{46}$$

This states that the change of phase across the junction is independent of time. The combination of (45) in (41) and (42) implies that

$$\frac{\partial \rho_1}{\partial t} = -\frac{\partial \rho_2}{\partial t} \tag{47}$$

This equation acts as a conservation equation, which states that the time rate of increase of the Cooper pair density in one superconductor equals the time rate of decrease of the Cooper pair density in the other superconductor. The magnitude of the pair current is proportional to the time rate of change of the density, $j_2 = \partial\rho_2/\partial t$. Hence we conclude that j is proportional to $\sin \Delta\theta$ and write finally that

$$j = j_0 \sin \Delta\theta \tag{48}$$

where j_0 is a function of the properties of the barrier, which includes the coupling term T and the temperature. This current density j_0 is the maximum dc supercurrent density that can pass through the junction and is called the critical current density.

The second fundamental relation for a Josephson junction is to note that a dc potential Φ is created across the junction if a current density j which is larger than j_0 passes through the junction. This can be understood in the following manner. Some of the current will be carried through the junction by single normal electrons that tunnel through the barrier. If the chemical potentials of the two superconductors are $2\mu_1$ and $2\mu_2$, the potential difference is given by

$$2q\Phi = 2\mu_1 - 2\mu_2$$

Let us use a convention that the potential is equal to zero at the midpoint of the oxide layer. Then (35) and (36) must be modified to read

$$i\hbar \frac{\partial\psi_1}{\partial t} = \hbar\Gamma\psi_2 + q\Phi\psi_1 \tag{49}$$

$$i\hbar \frac{\partial\psi_2}{\partial t} = \hbar\Gamma\psi_1 - q\Phi\psi_2 \tag{50}$$

The wave functions given by (37) and (38) are substituted into (49) and we obtain

$$i\hbar \left\{ \frac{\partial \sqrt{\rho_1}}{\partial t} + i\sqrt{\rho_1} \frac{\partial \theta_1}{\partial t} \right\} e^{i\theta_1} = \hbar \Gamma \sqrt{\rho_2} e^{i\theta_2} + q\Phi \sqrt{\rho_1} e^{i\theta_1} \tag{51}$$

Let (51) be multiplied by $(\rho_1)^{1/2} \varepsilon^{-i\theta_1}$ and make the substitution $\Delta\theta = \theta_2 - \theta_1$. As in the previous case where the chemical potentials were neglected, equate the real and the imaginary parts of the resulting equation and obtain

$$\frac{\partial \rho_1}{\partial t} = 2\Gamma \sqrt{\rho_1 \rho_2} \sin \Delta\theta \tag{52}$$

and

$$\frac{\partial \theta_1}{\partial t} = -\frac{q\Phi}{\hbar} - \Gamma \sqrt{\frac{\rho_2}{\rho_1}} \cos \Delta\theta \tag{53}$$

Note that (52) is the same as the dc Josephson effect case given in (41) and that an additional dc term $q\Phi/\hbar$ has entered into the description for the time rate of change of the phase. Similar equations are obtained following the same procedure for the wave function ψ_2, and we obtain

$$\frac{\partial \rho_2}{\partial t} = -2\Gamma \sqrt{\rho_1 \rho_2} \sin \Delta\theta \tag{54}$$

and

$$\frac{\partial \theta_2}{\partial t} = \frac{q\Phi}{\hbar} - \Gamma \sqrt{\frac{\rho_1}{\rho_2}} \cos \Delta\theta \tag{55}$$

If the Cooper pair densities are again set equal, we obtain

$$\frac{\partial \Delta\theta}{\partial t} = \frac{\partial(\theta_2 - \theta_1)}{\partial t} = \frac{2q\Phi}{\hbar} \tag{56}$$

This implies that the dc voltage Φ will create a time varying or oscillating change of phase. Equation (56) can be integrated since the term containing the dc voltage Φ on the right side is independent of time and the instantaneous phase difference can be obtained. It will increase linearly with time. Equation (56) also tells us that the supercurrent will oscillate in time as a function of the dc voltage Φ applied across the diode. The frequency ν of this oscillation is given by

$$2\pi\nu = \frac{2q\Phi}{\hbar} \tag{57}$$

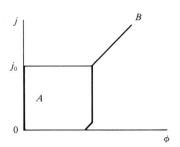

Figure 10.11 Schematic plot of the current through a Josephson diode as a function of the voltage across it. *A* is the pair current and *B* is due to the normal single-electron tunneling.

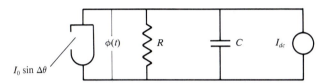

Figure 10.12 Equivalent circuit of a real Josephson junction with a current bias.

Using the numerical values for the constants, this corresponds to $483594\,\text{GHz/V}$. Since q and h are known to high accuracy, the Josephson junction is now the new dc voltage standard since we can accurately measure the oscillating frequency v.

In Figure 10.11, the current–voltage characteristics of a Josephson junction are displayed. Note that the voltage does not change even as the current density changes from zero to the value equal to j_0. In addition to the Josephson current–voltage relationship of the supercurrent of Cooper pairs, the current–voltage relation for the electron tunneling is also given. Note that when the critical current j_0 is exceeded there is a sudden jump from the dc pair current to the dc current due to the normal single electron tunneling. This is indicated with the horizontal line at j_0. The supercurrent through the junction oscillates in time with a frequency given by (57). If the current is decreased to zero from this value where an oscillation is observed, a hysteresis effect may be produced. In these curves, we have assumed that the external resistance R in the circuit (Figure 10.8) is large with respect to the diode.

Finally, we suggest an equivalent circuit for a real Josephson junction, which incorporates a current bias as shown in Figure 10.12.

Example 10.3 Write the differential equation that describes the Josephson junction depicted in Figure 10.12 and suggest a mechanical analog for it.

Answer: Using Kirchhoff's law, we write

$$I_{\text{dc}} = C\frac{d\phi}{dt} + \frac{\phi}{R} + I_0 \sin \Delta\theta$$

or, using (56),

$$\frac{\partial \, \Delta\theta}{\partial t} = \frac{2q\phi}{\hbar},$$

we write

$$I_{dc} = \frac{\hbar}{2q} C \frac{d^2 \, \Delta\theta}{dt^2} + \frac{\hbar}{2q} \frac{1}{R} \frac{d \, \Delta\theta}{dt} + I_0 \sin \Delta\theta$$

This suggests a mechanical circuit of the form as shown.

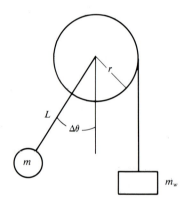

Newton's equation is written as

$$\text{Torque} = M_0 \frac{d^2 \, \Delta\theta}{dt^2}$$

where M_0 is the moment of inertia and the torque consists of three terms: (1) the applied torque $= m_w rg$, (2) the restoring torque due to gravity $= -mgL \sin \Delta\theta$, and (3) the opposing torque due to friction, which we write as $-Dd(\Delta\theta)/dt$. Therefore,

$$m_w rg = M_0 \frac{d^2 \, \Delta\theta}{dt^2} + D \frac{d \, \Delta\theta}{dt} + mgL \sin \Delta\theta$$

This mechanical model will be discussed further in Chapter 11.

It has been found that external high-frequency electromagnetic radiation can affect the characteristics of a Josephson junction. Therefore, it can be used as a detector of such radiation. With the bistable nature in the current–voltage characteristics (Figure 10.11), it has potential applications as a memory element. Finally, the Josephson effect can be used in a superconducting quantum interference device that carries the acronym SQUID. It is a very sensitive detector of changes in magnetic fields and as such, has been employed in geophysics, in detecting gravity waves, and in measuring variations associated with heart and brain activity in humans. It will be interesting to follow the industrial applications in the future.

V. Conclusion

As can be seen, several different solid-state effects have found application in various electronic circuits. These are based on different physical phenomena than the ordinary transistor that was described previously. In addition, the elements described here and in previous chapters can be incorporated into transmission-line circuits and we can propagate nonlinear waves such as solitons on them.

In the early months of 1987, we have been witnessing developments in the area of superconductivity that can only be thought of in the term "mind boggling." After tinkering with several oxide compounds instead of metallic alloys, Karl Alex Müller and Johannes Georg Bednorz were able to break through the $20°K$ temperature barrier by using various metallic oxides known as *ceramics* and achieved superconductivity at $35°K$. After this breakthrough, scientists from around the world entered the race to raise the temperature still higher. Paul C. W. Chu and Maw-Kuen Wu were the first to break the liquid nitrogen temperature barrier using mixtures of various rare earth compounds. This is a major accomplishment since superconductivity previously had been detected only at liquid helium temperatures (a few degrees kelvin) instead of these higher temperatures. Liquid helium temperatures make superconductivity mainly a scientific curiosity since liquified helium is expensive and difficult to handle. The use of liquid nitrogen as a coolant brings it into new realms since its price is of the order of the price of milk and is easily produced, transported, and stored. Scientific meetings where the latest results are presented turn into all-niters where the normal staid decorum is replaced by a "Woodstock-like" atmosphere with the scientists reaching for the microphone to announce a new compound or a temperature higher than $93°K$ which was reported by Chu and Wu.

Developments are occurring at such a rapid pace in this area that it is not possible to sit back and reflect on the technology in a text yet other than to alert the reader of its existence. Older scientists cannot recall this level of excitement in physics since the fissioning of the atom in the late thirties and the quantum mechanics development in the decade preceding that. The transistor and the laser certainly have stimulated the development of new industries and are worthy competitors for this "heart-throbbing" excitement but the thought of lossless storage, usage, and transmission of electrical power stimulates not only industries but also the wildest thoughts of futurologists. The Josephson junction may become central to the computer of tomorrow, medical instruments may become more sensitive, trains may levitate on magnetic fields, or magnets made from these materials may confine plasmas in fusion machines. The reader of these pages should keep abreast of the news that this community of engineers and scientists are bringing forth from their laboratories almost daily. The race to room-temperature superconductors continues as the dry-ice temperature barrier has recently been broken. New theories may be required.

PROBLEMS

1. Let the electron density pulse in an IMPATT diode be approximated as $n(x,t) = n_0(t)e^{-|x-L/2|}$. Calculate the current per unit area from this diode. You may assume that the diode is long.

2. An analysis of a Read diode (n^+pip^+ structure) leads to a small-signal terminal admittance of the form

$$y_{ac} = i\omega C \left\{ \frac{\omega\tau[1 - (\omega^2/\omega_a^2)]}{\omega\tau[1 - (\omega^2/\omega_a^2)] + i(1 - e^{-i\omega\tau})} \right\}$$

where

$$\omega_a = \sqrt{\frac{2(m + 1)J_{dc}}{\varepsilon E_A \tau_1}}$$

is the avalanche resonance frequency, J_{dc} is the dc bias current density, and τ_1 is the transit time through the avalanche region. The capacitance $C = \varepsilon A/L$. Using normalized values, $C = 1$, $\omega_a = 1$, and $\tau = 1$, show that the admittance will achieve its largest negative conductance at $\omega \approx 2$.

3. Plot the admittance of a Read diode as a function of frequency.

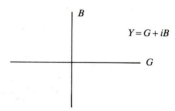

4. Let the distribution of electrons in the upper valley of a Gunn diode be governed by the Maxwell–Boltzmann distribution

$$\frac{n_U}{n_L} = \frac{N_U}{N_L} e^{-\Delta E/(k_B T)}$$

At room temperature, calculate this ratio if the upper band is $\Delta E \approx 0.3$ eV above the lower band. Discuss what this means in equilibrium conditions.

5. Calculate the RC time constant for a dielectric slab that has a conductivity σ and a dielectric constant ε.

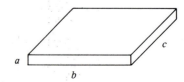

6. In analogy to the coupling of the Cooper pair wave functions across an oxide layer, show that a portion of an electromagnetic wave incident on an infinite planar dielectric at $x = 0$ whose thickness is Δx from the region $x < 0$ will propagate into the region $x > \Delta x$.

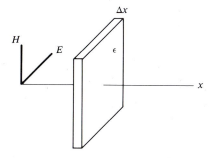

7. Show that (35) and (36) can be written as a second-order differential equation. Obtain the most general solution for one of the wave functions and sketch its behavior as a function of x.

8. Derive Equations (41) to (44).

9. Derive Equations (52) and (53).

10. For Example 10.3, explicitly state the equivalent terms between the Josephson junction and the mechanical analog. If possible, construct such a device.

11. From the two equations (48) and (56), show that we can model the Josephson junction with a nonlinear inductor $L(j)$, where $L(j) = L_0[\sin^{-1}(j/j_0)]/(j/j_0)$, where $L_0 = \hbar/(2qj_0)$.

12. Assume the Josephson junction is inserted in a circuit in parallel with a capacitor. Show that the resonant frequency will depend nonlinearly on the current in the junction using the results of Problem 11.

Chapter Eleven

Transmission-Line Applications
of Solid-State Devices

The reader of the words set on these pages has now probably reached a stage where the desire to apply the knowledge is burning deep. Rather than quash these embers, which have been nurtured in the previous pages or in earlier electronic circuit courses, we will keep the fires burning by suggesting applications of a nonconventional nature. The application of a solid-state device in a transmission-line milieu should stimulate the reader to study nonlinear phenomena with a little less trepidation. Using either a reverse-biased *pn* junction or a Josephson junction as an additional shunt element in a normal transmission line, the reader will be able to climb aboard the **Soliton** express, which just left the station in 1965 and is still boarding passengers who want to contribute their knowledge of electronic devices in order to assist in the explanation of this nonlinear wave. Our goal in this chapter is to help the reader with the loading of the baggage needed to make the transfer from the old linear steam engine to the modern nonlinear diesel-electric locomotive.

The technique that will be employed will be to describe the distributed transmission line, then derive the equation that models the line, and finally obtain the solution. At first glance, the mathematics may seem formidable and the concepts opaque; it is hoped that the reader will see the light at the end of the tunnel such that the train that has been boarded will arrive safely at the next station. Having boarded the Soliton express at the transmission-line station with its electronic components, we will then pass on to the next station, where the optical fiber with all its communication possibilities resides. It is convenient in the discussion that follows to identify the sections by using the titles of the nonlinear describing equation.

I. Korteweg–de Vries Equation

The first nonlinear wave equation that will be examined is the Korteweg–de Vries (KdV) equation, which was originally derived in the last decade of the nineteenth century to describe nonlinear waves that were observed to propagate in shallow channels of water for long distances. We will obtain a similar equation for a transmission line containing a varactor diode as the shunt element and show that pulses of a particular shape (called **solitons**) can propagate along this line. A consequence of this derivation is that the resulting soliton, which can survive a collision with another soliton, can be used to store or delay information that is encoded in the signal.

Let us examine the section of a transmission line shown in Figure 11.1. The units of the various elements are L (henries/length), C_N (farads/length), and C_s (farads-length). The shunt capacitor is a reverse-biased ***pn* junction diode**, which is also called a varactor diode (or "varicap") in Chapter 6.

The current through the nonlinear diode is given by

$$I_{\text{diode}} = \frac{\partial Q(\phi)}{\partial t}$$

where $Q(\phi)$ is the charge stored in the diode. The set of partial differential equations that describe the transmission line in the limit $\Delta x \to 0$ is given by

$$\frac{\partial I}{\partial x} + \frac{\partial Q(\phi)}{\partial t} = 0 \tag{1}$$

$$\frac{\partial \phi}{\partial x} + L\frac{\partial I'}{\partial t} = 0 \tag{2}$$

$$\frac{\partial \phi}{\partial x} + \frac{1}{C_s}\int (I - I')\, dt = 0 \tag{3}$$

In (2) and (3), I' is the current through the inductor and $I - I'$ is the current through the series capacitor C_s. The integral in (3) is removed by a differentiation with respect to time. These three equations are then combined by differentiating (3) with respect to x and t again, where we write

$$\frac{\partial^4 \phi}{\partial x^2\, \partial t^2} + \frac{1}{C_s}\left\{ \frac{\partial(\partial I/\partial x)}{\partial t} - \frac{\partial(\partial I'/\partial t)}{\partial x} \right\} = 0$$

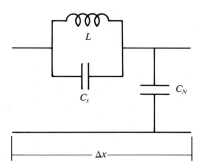

Figure 11.1 Section of a transmission line. The shunt capacitor is a reverse-biased varactor diode, and the series capacitor represents the capacitance between wires of the inductance L.

Employing (1) and (2) in this equation, we obtain

$$\frac{\partial^4 \phi}{\partial x^2 \, \partial t^2} + \frac{1}{C_s}\left[\frac{\partial\{-[\partial Q(\phi)]/\partial t\}}{\partial t} - \frac{\partial[-(1/L)(\partial\phi/\partial x)]}{\partial x}\right] = 0$$

or finally

$$\frac{\partial^4 \phi}{\partial x^2 \, \partial t^2} + \frac{1}{C_s L}\frac{\partial^2 \phi}{\partial x^2} - \frac{1}{C_s}\frac{\partial^2 Q(\phi)}{\partial t^2} = 0 \tag{4}$$

To carry the derivation further, an approximation must be made for the charge stored in the nonlinear diode. This charge is given by $Q(\phi) = \phi C(\phi)$, where $C(\phi)$ is the nonlinear capacitance of the diode. We could alternatively define the nonlinear capacitance from $C(\phi) = dQ(\phi)/d\phi$. The capacitance decreases as the voltage increases since the increased voltage increases the depletion width of the diode. Hence it is possible to approximate this charge as

$$Q(\phi) \approx C_0 \phi - C_1 \phi^2 \tag{5}$$

where C_0 is the linear capacitance and C_1 is a coefficient of the nonlinear contribution. Substituting (5) in (4), we obtain

$$\frac{C_s}{C_0}\frac{\partial^4 \phi}{\partial x^2 \, \partial \tau^2} + \frac{C_1}{C_0}\frac{\partial^2(\phi^2)}{\partial \tau^2} - \frac{\partial^2 \phi}{\partial \tau^2} + \frac{\partial^2 \phi}{\partial x^2} = 0 \tag{6}$$

where $\tau = t/(LC_0)^{1/2}$. This normalizes the time to the propagation velocity of the linear transmission line with the series capacitor $C_s \approx 0$.

In the limit of a linear line with no series capacitance (C_1 and C_s both equal to zero), (6) reduces to the second-order linear wave equation

$$-\frac{\partial^2 \phi}{\partial \tau^2} + \frac{\partial^2 \phi}{\partial x^2} = 0 \tag{7}$$

The solutions of (7) were obtained in Chapter 1 and can be written as

$$\phi = \phi_1(x - \tau) + \phi_2(x + \tau)$$

where ϕ_1 and ϕ_2 are the waves that propagate to increasing and decreasing values of position x as the time τ increases. Their shapes are determined by the initial excitation. Note that the velocity of the signal with this normalization is given by $1/(LC_0)^{1/2}$.

In the limit of the nonlinear term being equal to zero ($C_1 = 0$) but the series capacitance not equal to zero ($C_s \neq 0$), equation (6) reduces to

$$\frac{C_s}{C_0}\frac{\partial^4 \phi}{\partial x^2 \, \partial \tau^2} - \frac{\partial^2 \phi}{\partial \tau^2} + \frac{\partial^2 \phi}{\partial x^2} = 0 \tag{8}$$

To gain an understanding of the effects of including this additional term, let us assume that a wave of the form

$$\phi = \phi_0 e^{i(\omega \tau - kx)} \tag{9}$$

is excited and is propagating on the line. Substitute (9) into (8) and obtain

$$\frac{C_s}{C_0}(-ik)^2(i\omega)^2\phi - (i\omega)^2\phi + (-ik)^2\phi = 0$$

Note that ϕ is common to all three terms and it can be factored out from the equation. Hence we obtain an equation that is called a dispersion relation, which is given by

$$\omega^2 = \frac{k^2}{1 + (C_s/C_0)k^2}$$

Recall that the time has been normalized by the velocity $1/(LC_0)^{1/2}$. Therefore, we write

$$\omega = \frac{k}{\sqrt{LC_0 + LC_sk^2}} \approx \frac{1}{\sqrt{LC_0}}\left[k - \frac{C_s}{2C_0}k^3\right] \tag{10}$$

The approximation is valid for small values of k (or long wavelength since $k = 2\pi/\lambda$). A graph of this dispersion relation is shown in Figure 11.2. Note that at low frequencies ($\omega \to 0$) the phase velocity of the propagating wave, which is defined as $v_\phi = \omega/k$, is almost a constant. However, as the frequency increases, the phase velocity decreases. This will later set up a situation for the case when the nonlinear diodes are included, where there is a robust balance between harmonics that are created due to the nonlinearity of the diode and the dispersion that is found in the model of the linear transmission line. We will see that this is crucial for the formation of stable pulses that can form and propagate along the nonlinear dispersive transmission line.

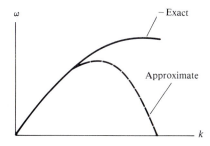

Figure 11.2 Dispersion curve corresponding to the linear transmission line given in Figure 11.1.

To derive the KdV equation, we will introduce and make use of a perturbation expansion, where a small expansion parameter μ is incorporated in the expansion. We will also make use of the fact that we are seeking stable propagating pulses, which are moving at almost the linear velocity $1/(LC_0)^{1/2}$ (which in the normalized units is 1). Therefore, let

$$\phi = \mu \phi^{(1)} + \mu^2 \phi^{(2)} + \cdots \tag{11}$$

$$\xi = \mu^{1/2}(x - \tau) \tag{12}$$

$$\eta = \mu^{3/2} \tau \tag{13}$$

We will assume that no large potential exists and the perturbation of the potential will be of first order in the expansion parameter μ. Since we are assuming that the pulse should be stable, it is reasonable to assert that the temporal variation (13) of the pulse is of higher order than the propagation characteristics that are determined by (12). The ordering of 1/2 and 3/2 also reflects the ordering of the approximate dispersion relation given in (10). Now employ the variables defined in (11) to (13) in (6). Recall from the chain rule that

$$\frac{\partial}{\partial \tau} = \frac{\partial}{\partial \xi}\left(\frac{\partial \xi}{\partial \tau}\right) + \frac{\partial}{\partial \eta}\left(\frac{\partial \eta}{\partial \tau}\right) = \frac{\partial}{\partial \xi}(-\mu^{1/2}) + \frac{\partial}{\partial \eta}(\mu^{3/2}), \quad \text{etc.}$$

Equating the terms that have the same power of the parameter μ, we find to lowest order the linear wave equation (7). To the next higher order, we obtain (for simplicity, let $\phi^{(1)} = \phi$)

$$\frac{\partial \phi}{\partial \eta} + \frac{C_1}{C_0} \phi \frac{\partial \phi}{\partial \xi} + \frac{C_s}{2C_0} \frac{\partial^3 \phi}{\partial \xi^3} = 0 \tag{14}$$

Equation (14) is the KdV equation that we are seeking. It was derived originally for studying water wave propagation along shallow channels, with particular emphasis on pulse propagation. It can be derived for waves in other media, a particular example being ion acoustic waves in a plasma. It was derived here for a transmission line consisting of inductors, capacitors, and varactor diodes. All three media have received considerable experimental and theoretical attention recently and new results are coming forth at a rapid pace.

Prior to obtaining a solution of the KdV equation, we will examine the solution of its linearized form. This will allow us to see how dispersion affects the propagation of a pulse that is governed by the equation

$$\frac{\partial \phi}{\partial \eta} + \frac{C_s}{2C_0} \frac{\partial^3 \phi}{\partial \xi^3} = 0 \tag{15}$$

Several techniques are available to obtain a solution of (15). We will apply the same technique that was used to solve the diffusion equation and which is described in Appendix A. Hence we will obtain a self-similar solution.

Following the procedure outlined there, we find the self-similar variables to be

$$\zeta = \frac{\xi}{\eta^{1/3}} \quad \text{and} \quad \Xi = \frac{\phi}{\eta^{-1/3}} \tag{16}$$

Substituting (16) into (15), we obtain

$$-\frac{1}{3}\zeta\frac{d\Xi}{d\zeta} + \left(-\frac{1}{3}\right)\Xi + \frac{C_s}{2C_0}\frac{d^3\Xi}{d\zeta^3} = 0 \tag{17}$$

In writing this equation, we are restricting our interest to the response to a very localized stimulus (say a δ function). The first integral of (17) is given by

$$-\frac{1}{3}\zeta\Xi + \frac{C_s}{2C_0}\frac{d^2\Xi}{d\zeta^2} = k_1 \tag{18}$$

where k_1 is a constant of integration that will be taken to be zero, since we are looking for a pulse solution.

Equation (18) can be recognized as Bessel's differential equation of order $\frac{1}{3}$. This is sometimes called the Airy differential equation, and its solution is given by

$$\Xi = \text{Ai}\left[\left(\frac{C_s}{6C_0}\right)^{1/3}\zeta\right] \tag{19}$$

A picture of an Airy function is shown in Figure 11.3. The high-frequency oscillations that trail the larger pulse have an increasing frequency or decreasing period between the zero crossings.

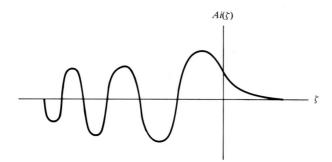

Ai(ζ)

ζ

Figure 11.3   Airy function.

Substitute the laboratory variables into (19) to obtain

$$\phi = \eta^{-1/3}\Xi = \eta^{-1/3}\,\text{Ai}\left[\left(\frac{C_s}{6C_0}\right)^{1/3}\frac{\xi}{\eta^{1/3}}\right]$$

$$= \tau^{-1/3}\,\text{Ai}\left[\left(\frac{C_s}{6C_0}\right)^{1/3}\frac{x-\tau}{\tau^{1/3}}\right] \tag{20}$$

It is now possible to state the expected behavior of the linear response to a narrow pulse excitation. The δ function pulse contains all frequencies from 0 to ∞. However, the higher-frequency components travel with a slower velocity due to dispersion and trail behind the lower-frequency components. This corresponds to the higher-frequency oscillations in Figure 11.3, which trail the main pulse. Hence the solution of the linearized form of the KdV equation (14) predicts that dispersion will cause a pulse to spread out, since the high-frequency components that cause it to have a sharper form cannot stay together with the lower-frequency components.

To obtain a pulse solution of the nonlinear KdV equation (14) that is propagating with a constant velocity without changing shape as it propagates (i.e., a wave of "permanent profile"), we will assume that the variables ξ and η can be combined into a wave variable θ as

$$\theta = \xi - \eta \tag{21}$$

If this is done, (14) transforms to the nonlinear ordinary differential equation

$$-\frac{d\phi}{d\theta} + \frac{C_1}{C_0}\phi\frac{d\phi}{d\theta} + \frac{C_s}{2C_0}\frac{d^3\phi}{d\theta^3} = 0 \tag{22}$$

This can be integrated once to

$$-\phi + \frac{C_1}{2C_0}\phi^2 + \frac{C_s}{2C_0}\frac{d^2\phi}{d\theta^2} = A \tag{23}$$

where A is a constant of integration. Since we are looking for a pulse solution whose value is zero when we are far from the center of the pulse, this constant will be zero. A second integral of (23) can be obtained by multiplying (23) by $(d\phi/d\theta)\,d\theta$ and integrating to obtain

$$-\frac{\phi^2}{2} + \frac{C_1}{6C_0}\phi^3 + \frac{C_s}{4C_0}\left(\frac{d\phi}{d\theta}\right)^2 = B \tag{24}$$

where again the constant of integration B is set equal to zero due to our interest in obtaining a pulse-type solution.

The integral of (24) is given by

$$\phi = \phi_0 \, \text{sech}^2 \left\{ \left(\frac{C_1 \phi_0}{6 C_s} \right)^{1/2} \left[x - \left(1 + \left[\frac{C_1 \phi_0}{3 C_0} \right] \frac{t}{\sqrt{L C_0}} \right) \right] \right\} \qquad (25)$$

where ϕ_0 is the amplitude of the pulse and (25) has been written in the coordinates of the transmission line. The solution given in (25) is sometimes called a solitary wave since it has the shape of a pulse. In his original observation in 1834 of a water wave excited ahead of a moving barge, John Scott Russell called it the "great wave of translation." This wave was excited by the barge, which had rapidly stopped, and was found to propagate on a canal in Scotland. The descriptive report to the British Association for the Advancement of Science by Russell, who followed the wave on horseback for several miles, makes for delightful reading.

Certain general properties of the pulse solution given by (25) are apparent. First, the normalized velocity of the pulse

$$\text{velocity} = 1 + \left[\frac{C_1 \phi_0}{3 C_0} \right]$$

depends on the amplitude of the pulse. This means that a pulse that is launched at a point, say at $x = -5$, could catch up with a second pulse launched at the same time at $x = 0$ at some distance $x > 0$ if the amplitude of the first pulse is greater than that of the second. Also, the product of the pulse amplitude and the square of its width is a constant:

$$(\phi_0) \left(\sqrt{\frac{C_1 \phi_0}{6 C_s}} \right)^{-2} = \frac{6 C_s}{C_1}$$

A third property that arises out of the pulse solution of the nonlinear KdV equation is that the effects of the nonlinear term have balanced the dispersive effects such that a stable pulse propagates along the line.

A fourth property that is not apparent from the solution given in (25) is that if two of these pulses collide, either during the collision during the catching-up process or during a head-on collision, they will be unaffected except for a possible shift in phase in their location. They will, however, preserve their shape. This particlelike behavior led N. Zabusky and M. Kruskal to coin the word "soliton" in 1965 following a numerical investigation of the KdV equation in order to distinguish this solution of the KdV equation that has this peculiar collision property from an ordinary pulse solution. It has often been used as a definition for a soliton. Soliton properties are found in a fairly large class of nonlinear partial differential wave equations, and they carry the generic title "nonlinear evolution equations." A practical application for the soliton has been suggested in that information could be encoded into the soliton and this secure mode of propagation could only be decoded by someone who knew the proper "soliton combination"; it could therefore safely pass through enemy lines.

Our first application of a solid-state device in a transmission-line circuit has led to an experimental device that can form and allow the propagation of solitons. Considerable knowledge is currently being gained about this nonlinear wave in several areas of mathematics, science, and engineering. A second type of soliton will be described in the next section.

II. Sine–Gordon Equation

As a second example of the application of a solid-state device in a nonlinear transmission line, we will examine a transmission line that contains a Josephson junction. A section of this transmission line is shown in Figure 11.4.

The Josephson junction is defined by the set of equations

$$I = I_0 \sin \Delta\theta \tag{26}$$

and

$$\frac{\partial(\Delta\theta)}{\partial t} = \frac{2q\phi}{\hbar} \tag{27}$$

Both of these equations were derived in the previous chapter in the discussion leading to the physical mechanism of the Josephson junction.

To derive the transmission-line equation that incorporates a Josephson junction, we start with Kirchhoff's equations and write

$$\frac{\partial i}{\partial x} = C\frac{\partial \phi}{\partial t} + I_0 \sin \Delta\theta \tag{28}$$

and

$$\frac{\partial \phi}{\partial x} = L\frac{\partial i}{\partial t} \tag{29}$$

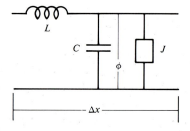

Figure 11.4   Transmission-line model of a Josephson junction strip line. Element J is the Josephson junction.

Differentiate (28) with respect to time and write

$$\frac{\partial(\partial i/\partial t)}{\partial x} = C\frac{\partial^2\phi}{\partial t^2} + I_0\frac{\partial(\sin\Delta\theta)}{\partial t}$$

Using (29), we write

$$\frac{1}{L}\frac{\partial^2\phi}{\partial x^2} = C\frac{\partial^2\phi}{\partial t^2} + I_0\frac{\partial(\sin\Delta\theta)}{\partial t}$$

Substitute (27) in this equation to eliminate the voltage ϕ:

$$\frac{1}{L}\frac{\hbar}{2q}\frac{\partial^3\Delta\theta}{\partial t\,\partial x^2} = C\frac{\hbar}{2q}\frac{\partial^3\Delta\theta}{\partial t^3} + I_0\frac{\partial(\sin\Delta\theta)}{\partial t}$$

This can be integrated once with respect to time to obtain

$$\frac{1}{L}\frac{\hbar}{2q}\frac{\partial^2\Delta\theta}{\partial x^2} - C\frac{\hbar}{2q}\frac{\partial^2\Delta\theta}{\partial t^2} = I_0\sin\Delta\theta \qquad (30)$$

where the constant of integration is set equal to zero. Equation (30) is called the sine–Gordon equation. This appellation follows from the name Klein–Gordon equation, where the term $\sin\Delta\theta$ is just $\Delta\theta$. This equation can be put in a normalized form:

$$\frac{\partial^2\Delta\theta}{\partial\chi^2} - \frac{\partial^2\Delta\theta}{\partial\tau^2} = \sin\Delta\theta \qquad (31)$$

where

$$\chi = \frac{x}{\sqrt{(1/L)(\hbar/2q)(1/I_0)}}$$

is normalized by the Josephson penetration length and

$$\chi = \frac{t}{\sqrt{C(\hbar/2q)(1/I_0)}}$$

is normalized by the Josephson plasma frequency.

Equation (31) has a mechanical analog that was brought out in Example 10.3, which should lead to some understanding of the predictions. Imagine a long rubber band that is stretched between two fixed points. Inserted in one side of the band and with equal separations is a set of pins. Due to gravity, the heads of the pins would all be close to the ground, as shown in Figure 11.5. If the pins at one end were rotated about the equilibrium value, it would couple to the next one and cause it to rotate. A "wave" of rotation would progress down the line. This would also be true even if the rotation were through the full 360°. This is similar to the propagation of a wave along a Josephson junction transmission line.

To obtain a solution of the sine–Gordon equation, we again make the plane wave assumption of letting

$$\xi = \chi - u\tau \tag{32}$$

where u is a constant velocity. The sine–Gordon equation is transformed to

$$\frac{d^2 \Delta\theta}{d\xi^2} = \frac{\sin \Delta\theta}{1 - u^2} \tag{33}$$

Equation (33) has solutions that are in terms of Jacobian elliptic functions. One such solution is indicated in the mechanical line shown in Figure 11.6. As these functions are usually not in the reader's normal handbook of readily used functions, we present one limiting case as

$$\Delta\theta = 4\tan^{-1}\left\{ \exp\left[\pm \frac{\xi - \xi_0}{1 - u^2} \right] \right\} \tag{34}$$

The function given in (34) is sometimes called a "gudermannian." It is also the soliton solution of the sine–Gordon equation.

If a Josephson junction had been used in this transmission line and the mathematics had been worked out in full detail, we would have found once again that a soliton would have been uncovered. Since this is operating at almost absolute zero degrees in temperature and we are talking about superconductivity, the reader can see that extremely low levels of power will be dissipated. Also, the motion of these "rotations" can be extremely fast, so this device may be useful for high-frequency microwave generation or high-speed computation.

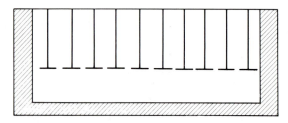

Figure 11.5   Mechanical analog of a Josephson junction transmission line with no applied torque.

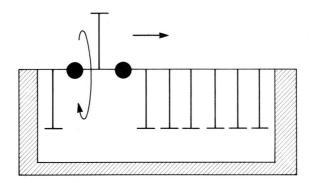

Figure 11.6   Mechanical analog solution of the sine–Gordon equation.

III. Nonlinear Schrödinger Equation

The third of the major nonlinear evolution equations that admits soliton solutions is the nonlinear Schrödinger equation or, as it is more commonly referred to, the NLS equation. From the name NLS equation, the reader can postulate that the equation should be related to the linear Schrödinger equation that was introduced in Chapter 1, where the properties of the solution were determined by a potential well U that would model the physical situation. The one-dimensional *linear* Schrödinger equation was written as

$$-i\hbar\frac{\partial\psi}{\partial t} = \left(\frac{\hbar^2}{2m}\right)\frac{\partial^2\psi}{\partial x^2} - U\psi \tag{35}$$

In the derivation of (35), we freely interchanged time and space derivatives,

$$-i\omega \to \frac{\partial}{\partial t} \quad \text{and} \quad ik \to \frac{\partial}{\partial x}$$

with the appropriate terms that were found in an energy relation. The same procedure will be employed here except that we will start with a slightly more general dispersion relation for the wave that is propagating in the media, which for example could be an optical glass fiber.

$$\omega = \omega(k, |\phi|^2) \tag{36}$$

Equation (36) states that the dispersion relation for the media depends on the *intensity* of the wave, since the intensity is proportional to $|\phi|^2$. In the discussion leading to the KdV equation, we have already encountered a nonlinear varicap whose value of capacitance depends on the amplitude of the applied voltage ϕ.

The wave $\psi(x, t)$ that is propagating in the fiber has a slow amplitude modulation and is written as

$$\psi(x, t) = \phi(x, t)e^{i(kx - \omega t)} \tag{37}$$

The frequency and the wave number of the modulation $\phi(x, t)$ are much smaller than the terms in the exponential term. If we take a time average of a quantity proportional to the intensity

$$\int_0^{2\pi/\omega} \psi\psi^* \, dt$$

we find that the time average is just $|\phi|^2$, which has not changed very much on the fast time scale contained in the exponential term. This term is sometimes called the Miller force or the ponderomotive force. Its derivation requires the following physical arguments. If a charged particle is oscillating in a slowly varying spatially inhomogeneous electric field, it will receive more of a "kick" where the field is stronger than where it is weaker. Since the field is spatially inhomogeneous, this will cause the particle to slowly drift in one direction. Its consequences are important, as will be shown next.

Let us expand (37) with a Taylor expansion about the point

$$k = k_0, \qquad \omega = \omega_0, \qquad \text{and } |\phi|^2 = |\phi_0|^2$$

Hence we write

$$\omega - \omega_0 \approx \left.\frac{\partial \omega}{\partial k}\right|_{k_0} (k - k_0) + \frac{1}{2} \left.\frac{\partial^2 \omega}{\partial k^2}\right|_{k_0} (k - k_0)^2 + \left.\frac{\partial \omega}{\partial |\phi|^2}\right|_{|\phi_0|^2} (|\phi|^2 - |\phi_0|^2) \tag{38}$$

In (38), we again make the substitution $-i(\omega - \omega_0) \to \partial/\partial t$ and $i(k - k_0) \to \partial/\partial x$. This leads to the following equation, where we again consider that the dispersion relation is *operating* on the wave ϕ.

$$i\left[\frac{\partial \phi}{\partial t} + \left.\frac{\partial \omega}{\partial k}\right|_{k_0} \frac{\partial \phi}{\partial x}\right] + \frac{1}{2} \left.\frac{\partial^2 \omega}{\partial k^2}\right|_{k_0} \frac{\partial^2 \phi}{\partial x^2} - \left.\frac{\partial \omega}{\partial |\phi|^2}\right|_{|\phi_0|^2} [|\phi|^2 - |\phi_0|^2]\phi = 0 \tag{39}$$

It is possible to simplify (39) by first noting that the two first-order derivative terms can be combined by making the substitution that

$$\xi = x - \left.\frac{\partial \omega}{\partial k}\right|_{k_0} t \quad \text{and} \quad \tau = t$$

This states that the modulation is moving with the group velocity. Hence (39) becomes

$$i\frac{\partial\phi}{\partial\tau} + a\frac{\partial^2\phi}{\partial\xi^2} + b|\phi|^2\phi = 0 \tag{40}$$

where

$$a = \frac{1}{2}\frac{\partial^2\omega}{\partial k^2}\bigg|_{k_0} \quad \text{and} \quad b = -\frac{\partial\omega}{\partial|\phi|^2}\bigg|_{|\phi_0|^2}$$

and $|\phi_0|^2 \approx 0$.

Comparing (35) and (40), we recognize that the potential U has been replaced with the nonlinear term that is proportional to the intensity $|\phi|^2$. Equation (40) is the NLS equation that we seek. Its solution can be effected by assuming it to be of the form

$$\phi(\xi\tau) = e^{i\Omega\tau}\Gamma(\xi) \tag{41}$$

Equation (40) becomes

$$-\Omega\Gamma + a\frac{d^2\Gamma}{\delta\xi^2} + b\Gamma^3 = 0 \tag{42}$$

This can be integrated twice if a "pulse-type" solution is assumed in that the perturbation and all its derivatives smoothly go to zero as $\xi \to \pm\infty$. The first integral is

$$-\Omega\frac{\Gamma^2}{2} + \frac{a}{2}\left(\frac{d\Gamma}{d\xi}\right)^2 + b\frac{\Gamma^2}{4} = \text{constant}$$

where the constant of integration is set equal to zero. This can be integrated

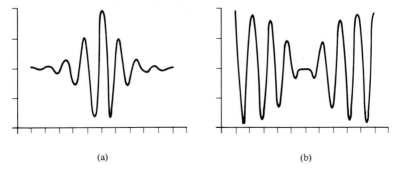

(a)   (b)

Figure 11.7   Two possible NLS solitons. (a) A "light" soliton. (b) A "dark" soliton.

again, and the solution is written as

$$\phi(\xi, \tau) = \sqrt{\frac{2\,\Omega}{b}}\,\mathrm{sech}\left[\sqrt{\frac{\Omega\,a}{b}}\,\xi\right]\exp\left(\frac{i\,\Omega}{b}\,\tau\right) \tag{43}$$

Recall that this is the amplitude modulation of a high-frequency wave. It can either be "light" or "dark," since the term $|\phi|^2$ is independent of sign. It propagates with the group velocity. A sketch of an NLS soliton is shown in Figure 11.7.

IV. Conclusion

In this chapter, we have suggested two applications of solid-state devices that are somewhat unconventional. In both cases, transmission lines permitted the propagation of solitons. The nonlinear wave equations that were derived for the resulting transmission lines are members of a large family of equations that are called nonlinear evolution equations. A third transmission line that did not rely on the solid-state element is the optical fiber and it was described by a third member of this family. These three equations are receiving considerable attention among mathematicians, and the soliton phenomenon is also receiving considerable attention among experimentalists of many persuasions. For example, solitons exist in plasmas and are found along many beaches, in the depths of oceans, and as the Giant Red Spot on Jupiter. It is not surprising, therefore, that electrical transmission lines made with the appropriate elements would also allow the existence and propagation of these entities. The reader is encouraged to explore this unique nonlinear phenomenon further as there remain several mysteries.

PROBLEMS

1. Show that Equation (5) is a reasonable approximation for a reverse-biased *pn* diode. State any approximations.

2. Show that $\phi = \phi_0 \sin(x - \tau)$ is a solution of Equation (7). Accurately sketch the wave as a function of position x at various times τ.

3. The velocity of a small amplitude wave that propagates along the line depicted in Figure 11.1 is given by $v = (LC_0)^{-1/2}$. Show that this is dimensionally correct.

4. In Chapter 1, it was noted that the dispersion relation, when multiplied with a wave function ψ, was considered to be operating on the wave function ψ. Show that the dispersion relation given in (10), $\omega \approx (1/LC_0)^{1/2}[k - (C_S/2C_0)k^3]$, leads to the linear KdV equation (15).

5. It is possible to construct a transmission line consisting of distributed capacitors C_S and diodes C_N and connect them to a wire that is loosely wound around a 3-cm-diameter rod. Typical values for the elements are $L\,\Delta X \approx 0.2\ \mu\mathrm{H}$, $C_S/\Delta X \approx 220$ pf, and $C_N\,\Delta X|_{\phi=0} \approx 500$ pf, where $\Delta X \approx 2$ cm.

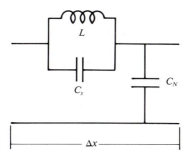

Using these values, find the cutoff frequency above which the line will not allow the propagation of a wave. Find the maximum wave number k that will be allowed in this case. Estimate the velocity of a wave in the nondispersive region of propagation.

6. Construct the experimental nonlinear transmission line described in Problem 5 and study the waves that can propagate along it.

7. For very small amplitudes $\Delta\theta$, the term $\sin\Delta\theta$ can be replaced by $\Delta\theta$. If this were done, obtain the dispersion relation for the linear sine–Gordon equation. Plot this dispersion relation.

8. Show that the Josephson plasma frequency and Josephson penetration length as defined following Equation (31) have the proper dimensions of $(\text{time})^{-1}$ and length.

9. Construct the mechanical analog of the Josephson junction transmission line using a rubber band and stick pins. Study the wave that can propagate along the line.

10. With the line constructed in Problem 9, try to collide two solitons by launching one from each end of the rubber band.

11. Using (32), sketch the expected soliton solution given in (34) and compare this prediction with your experimental line given in Problem 9.

Chapter Twelve

Charged-Particle Ballistics

The solemnity that we possessed when entering the archives and laboratories of the solid-state revolutionaries has been well rewarded with the technological revolution that has occurred in the last third of this century. It is perhaps time now to reflect on a slightly different topic, which falls under the general rubric of electrical ballistic devices, that has had an almost similar impact over a slightly longer period of time. The topics that will be described in the last portion of this testament have a foundation that does not rely on solid-state physics, although the reader is cautioned not to seal the early chapters since we may find it profitable later to return to them at critical stages. Hi-tech industry has found that charged-particle ballistic machines are useful in solid-state device manufacture through such processes as sputtering and molecular beam epitaxy. Contemporary newspaper accounts describe the "ballistic transistor" as the ultimate high-speed transistor that is just emerging from the research laboratory. The foundation of these devices is based on individual charged-particle motion in electric and magnetic fields and the collective interaction that can ensue due to the presence of neighboring charged particles in the system. This latter system has sometimes been called the fourth state of matter. These four states are solid, liquid, gas, and **plasma**.

An introduction to the plasma state is important since over 99.4% of the universe has decided to be in this ionized state. As the early twentieth-century scientists walked in the woods contemplating the ramifications of quantum mechanics, a few probably looked up into the skies and wondered what was up there. They hoped to answer the children's song, "Twinkle, twinkle little star. How I wonder what you are ...." The study of this new state of matter has many important consequences as the energy reservoirs of the earth are depleted, new processing technologies are developed, and space travel and communication become commonplace. We will also be able to answer the child's question, which was put to song long ago.

I. Charged-Particle Motion in an Electric Field

The introduction of the topic of charged-particle motion in a region where both electric and magnetic fields can act on it is most easily facilitated by considering them separately and neglecting the contribution of the gravitational force.

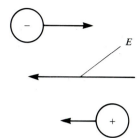

Figure 12.1   Charged-particle
motion in an electric field.

Newton's equation of motion for the charged particle is written for electrons as

$$m_e \frac{d\mathbf{v_e}}{dt} = -q\mathbf{E} \tag{1}$$

and for ions as

$$M_i \frac{d\mathbf{v_i}}{dt} = q\mathbf{E} \tag{2}$$

Note that the force is a vector and that the ions and the electrons will be accelerated in the opposite directions, as shown in Figure 12.1.

  If the electric field had been created by an external battery connected between a grid and a cathode where the charged particles (electrons in this case) are created, the particles would experience an acceleration only in the region $0 \le x \le L$, as shown in Figure 12.2. We will assume that the cathode is at $x = 0$ and the grid is at $x = L$, so the electrons will be accelerated to the right. The electrons are "falling down a potential hill" as they approach the region whose potential is higher.

  Since this can be considered to be a one-dimensional system, we can neglect any transverse components of the electric field and integrate (1) as

$$m_e \frac{dv_x}{dt} = qE \tag{3}$$

The minus sign has disappeared since the electric field and the acceleration have

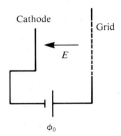

Figure 12.2   Electron gun for
accelerating electrons to a velocity
given by the applied potential.

the opposite direction. The integral of (3) is

$$m_e v_x = qEt + A \tag{4}$$

where A is a constant of integration. If the electrons are at the cathode ($x = 0$) at $t = 0$ and they have zero velocity there, this constant A will equal zero. The explanation of how the electrons actually leave the cathode will be given in the next chapter, where the Richardson–Dushmann equation is derived.

The second integral of (4) yields

$$m_e x = \frac{qEt^2}{2} + B \tag{5}$$

where B is another constant of integration. This constant will also be zero since the electrons are assumed to be at the cathode ($x = 0$) at $t = 0$. The transit time ΔT for the electrons from the cathode to the anode can be computed from (5) as

$$m_e L = \frac{qE(\Delta T)^2}{2} \tag{6}$$

From (4), the velocity of the electrons as they leave the grid region is given by

$$v_x = \frac{qE}{m_e} \Delta T \tag{7}$$

Eliminate the transit time ΔT between (6) and (7) and write

$$v_x = \frac{qE}{m_e} \sqrt{\frac{2m_e L}{qE}}$$

or

$$v_x = \sqrt{\frac{2qEL}{m_e}}$$

Since the electric field is assumed to exist only between the cathode and the grid, we may approximate its value as $E \approx \phi_0/L$ and write this velocity as

$$v_x = \sqrt{\frac{2q\phi_0}{m_e}} \tag{8}$$

It is possible to interpret this result from an energy consideration as follows. The electron as it passes through the electric field will gain kinetic energy; the incremental increase is

$$\Delta(\mathrm{KE}) = \frac{1}{2} m_e v_x^2$$

This increase must equal the potential energy that it lost while passing from the cathode, where its potential energy is $q\phi_0$, to the grid, where its potential energy is zero.

$$\Delta(\text{PE}) = q\phi_0$$

Equating these two energies and solving for the velocity v_x, we again obtain (8).

Example 12.1   Show formally that the energy gained by the electron is equal to the difference of potentials between the cathode and the grid.

Answer: The energy that is expended to accelerate the electron is given by the integral $\int_0^L F\,dx$. Apply this to both sides of the equation of motion (3) and write

$$\int_0^L m_e \frac{dv_x}{dt}\,dx = \int_0^L qE\,dx$$

The integral on the left side can be written as

$$\int_0^{\Delta T} m_e \frac{dv_x}{dt}\frac{dx}{dt}\,dt = \int_0^{v_x} m_e v_x\,dv_x = \frac{1}{2}m_e v_x^2$$

The integral on the right side can be written as

$$\int_0^L q\left(\frac{d\phi}{dx}\right)dx = \int_0^{\phi_0} q\,d\phi = q\phi_0$$

where the sign of the charge has canceled the sign in $E = -d\phi/dx$. These two expressions are the kinetic energy gained by the electron at the expense of the potential energy.

The preceding electron gun has a limited use as it stands since it can be directed in only one direction. Physically picking it up and pointing it in a different direction is not considered to be useful. It is, however, possible to deflect the electrons electronically using the arrangement shown in Figure 12.3.

The electrons that leave the electron gun have a constant velocity equal to v_x, which is given by equation (8). There is no force on these electrons in the x direction in the region to the right of the grid. However, as they approach the region that is enclosed within the parallel plates, they will be deflected in the y direction by the electric field created by the voltage $\Delta\phi$. This deflection is computed from Newton's equation of motion:

$$m_e \frac{dv_y}{dt} = qE = q\frac{\Delta\phi}{D} \qquad (9)$$

where again all fringing fields have been neglected. If we choose the midpoint

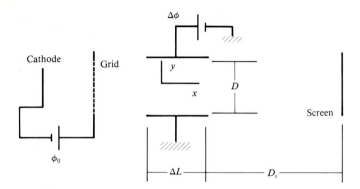

Figure 12.3   Electron gun with deflecting plates.

between the plates to have the coordinate $y = 0$, then the first integral of (9) yields

$$m_e v_y = \frac{q \, \Delta\phi}{D} t + C \tag{10}$$

and the second integral becomes

$$m_e v_y = \frac{q \, \Delta\phi}{D} \frac{t^2}{2} + Ct + D \tag{11}$$

where the constants of integration C and D will be set equal to zero, since the electrons are assumed to be just along the midplane between the plates as they enter the deflecting plate region. It is possible to estimate the time $\delta\tau$ that the electrons will spend between the deflecting plates from

$$\delta\tau = \frac{\Delta L}{v_x} \tag{12}$$

Hence the actual deflection in the y direction as the electrons leave the region of the deflecting plates can be computed as

$$y = \frac{q}{2m_e} \frac{\Delta\phi}{d} (\delta\tau)^2 \tag{13}$$

Its velocity in the y direction is

$$v_y = \frac{q}{m_e} \frac{\Delta\phi}{d} \delta\tau \tag{14}$$

Beyond the ends of the deflecting plates, there will be no force acting on these electrons and they will "free-stream." At a location D_s beyond the end of the deflecting plates, a screen is located. The electrons will "hit" the screen at a location Y

$$Y = y_{\text{end of deflecting plates}} + v_y \,(\text{time of flight})$$

$$= \frac{q}{2m_e} \frac{\Delta\phi}{d} (\delta\tau)^2 + \left(\frac{q}{m_e} \frac{\Delta\phi}{d} (\delta\tau) \right) \frac{D_s}{v_x}$$

$$= \frac{q}{2m_e} \frac{\Delta\phi}{d} \left(\frac{\Delta L}{v_x} \right)^2 + \left(\frac{q}{m_e} \frac{\Delta\phi}{d} \left(\frac{\Delta L}{v_x} \right) \right) \frac{D_s}{v_x}$$

or, finally,

$$Y \approx \frac{\Delta L}{d} \frac{\Delta\phi}{2\phi_0} D_s \tag{15}$$

where (8) has been employed and the final deflection Y is assumed to be much larger than its value at the edge of the deflecting plates. Note that for a given electron accelerating gun voltage, the deflection Y is directly proportional to the deflecting voltage $\Delta\phi$.

It is obvious that the beam could also be deflected in the other transverse direction and a two-dimensional array could be created. A tube based on this effect is called a cathode-ray tube (CRT) since the deflected electrons may possess sufficient energy to emit a photon at a precise location on a phospor-coated screen; hence visual effects can result.

The extension of this analysis to the study of an ion gun with positive ions just involves the changing of the polarity of the appropriate voltages. The temporal response will be much slower due to the mass difference.

II. Charged-Particle Motion in a Magnetic Field

The motion of charged particles in a magnetic field is also described by Newton's equation of motion for electrons as

$$m_e \frac{d\mathbf{v_e}}{dt} = -q(\mathbf{v_e} \times \mathbf{B}) \tag{16}$$

and for ions as

$$M_i \frac{d\mathbf{v_i}}{dt} = q(\mathbf{v_i} \times \mathbf{B}) \tag{17}$$

The force is again a vector; in fact, the force will be in a direction that is perpendicular to both the direction of the velocity and the magnetic field. The ions and electrons will be accelerated in different directions due to the different signs of the charge as in the case of the electric field. The major difference for the case of the magnetic field is that the particles must be in motion for a force to exist. Also, the magnetic field will exert no force on the particles if their velocity is in the direction of the magnetic field. The motion of the electrons and the ions is indicated in Figure 12.4.

Note in Figure 12.4 that the radius of curvature depends on the mass of the particle. This can be shown as follows. Since the ions are in uniform circular motion, they will experience a centrifugal force whose magnitude is given by

$$M_i a = \frac{M_i v_i^2}{r}$$

where r is the radius of curvature and a is the acceleration. If we assume that the vector directions are properly included, the magnitude of the force in (17) is given by

$$q v_i B$$

Equating these two expressions for the force, we obtain

$$r_L = \frac{M_i v_i}{qB} \tag{18}$$

We now make an analogy with a wave that has a wavelength defined from $\lambda = c/f_c$, where c is the velocity of the wave and f_c is its frequency. We let this wavelength be equal to the circumference of the circle defined by its radius of curvature r_L and obtain

$$2\pi r_L = \lambda = \frac{v_i}{f_c}$$

or

$$r_L = \frac{v_i}{\omega_c} \tag{19}$$

where r_L is called the Larmor radius and ω_c is the angular frequency of rotation.

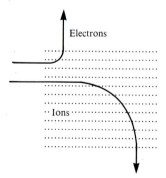

Electrons

Ions

Figure 12.4   Trajectories of electrons and ions into a localized magnetic field that is pointed out of the page.

Example 12.2   Calculate the energy expended in having a charged particle pass through a uniform magnetic field.

Answer: The work that is performed on the particle is given by the expression

$$\Delta W = \int_a^b \mathbf{F} \cdot \mathbf{dr}$$

This equation will equal zero since the force on the particle and the distance that the particle travels are perpendicular to each other.

The mass dependence of the Larmor radius has some interesting and practical consequences that have received attention. Assume that we wish to diagnose a group of unknown ions and determine their composition. A common technique is to shoot the ions into a uniform magnetic field region and detect them in various bins whose locations are determined by their Larmor radii, as shown in Figure 12.5. This concept also has practical implications in purifying materials.

That the charged particles are "confined" by a magnetic field has some very important ramifications for the study of plasma physics in the next chapter. If we wanted to create a large volume containing almost an equal number of ions and electrons and have this volume of plasma remain in the same location in space, it would have to be confined and the losses to the walls must be reduced. This can be done using magnetic fields, and we illustrate one possible technique in Figure 12.6. Let the surface area of the container be A. In Figure 12.6(a), the particles are free to escape to any part of the wall and the loss of these charged particles is proportional to A. In Figure 12.6(b), the charged particles cannot cross the magnetic field as given in (16) and (17) and as shown in Figure 12.4. They are free, however, to move along the magnetic field lines to the wall. But the field lines are normal to the wall at a much smaller area ΔA. Therefore, the loss is reduced in Figure 12.6(b) by the ratio $\Delta A/A$. This particular device illustrates the idea of a magnetically confining structure.

The earth and several other planets have a naturally occurring confining magnetic structure that confines charged particles that emanate from the sun in what is called the solar wind. Satellites have probed these regions of charged particles. The structure depicted in Figure 12.7 around the earth is called the Van

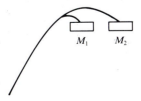

Figure 12.5   Mass spectrometer. A uniform magnetic field is out of the paper and the mass of ion 2 is greater than 1.

M_1 M_2

Magnets

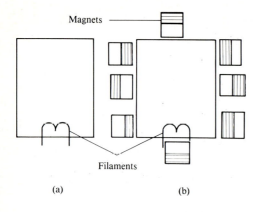

Filaments

(a)   (b)

Figure 12.6 Plasma confinement device. (a) No surface magnetic field. (b) Multidipole surface magnetic field. The shaded region corresponds to the same pole face.

N

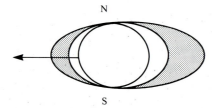

S

Figure 12.7 Regions of charged particles captured by the earth's magnetic field as it sweeps through the solar wind.

Allen belt in honor of its discoverer. There are several other magnetic "bottles" for the confinement of a plasma. Some of them will be described in the next chapter.

III. Charged-Particle Motion in an Electric and Magnetic Field

To predict the behavior of charged particles in a region where there is both an electric and a magnetic field, we just have to add the contributions of the two forces and Newton's equations become

$$m_e \frac{d\mathbf{v_e}}{dt} = -q(\mathbf{E} + \mathbf{v_e} \times \mathbf{B}) \tag{20}$$

and

$$M_i \frac{d\mathbf{v_i}}{dt} = q(\mathbf{E} + \mathbf{v_i} \times \mathbf{B}) \tag{21}$$

One interesting consequence of this inclusion of both fields is that there exists a steady-state drift of the particles that is independent of their charge or mass.

From (20) or (21) with $d/dt = 0$, we write

$$\mathbf{E} + \mathbf{v} \times \mathbf{B} = 0$$

Take the cross product of this equation with the magnetic field **B** and write

$$\mathbf{E} \times \mathbf{B} + (\mathbf{v} \times \mathbf{B}) \times \mathbf{B} = \mathbf{E} \times \mathbf{B} + [(\mathbf{v} \cdot \mathbf{B})\mathbf{B} - (\mathbf{B} \cdot \mathbf{B})\mathbf{v}] = 0$$

The vector direction associated with the second term is in the direction of the magnetic field and hence it is perpendicular to the first term, which is perpendicular to the magnetic field due to the cross product. The third term has a component of **v** that is in the direction of the first term, and we find a "guiding center drift" of particles, which is given by

$$\mathbf{v} = \frac{\mathbf{E} \times \mathbf{B}}{B^2} \tag{22}$$

If a charged particle is injected into such a cross-field system with this velocity, it will travel in a straight line.

IV. Elastic Collisions

The subject of collisions was touched on in Chapter 3 when we described the evolution of nonequilibrium statistics. Here we will provide more details and a further interpretation. Collisions between particles can be divided into two classes based on their state after the collision. If the particles are unaltered because of the collision, we refer to the collision as an *elastic* collision and this will be examined first. If their final state is different than it was before the collision, it is called an *inelastic* collision; fusion, ionization, excitation, dissociation, and the like would be examples of this type.

It is possible to treat an elastic collision using quantum mechanics or to treat it classically. The latter approach is not strictly correct when the dimensions are on the order of atomic dimensions, but it is more elementary for presentation. Also, it is valid if the discrete energy levels can be replaced by their continuum distribution. A rough estimate of the range of validity of the classical calculation is that the square of the wavelength of the impinging particle is much less than the calculated cross section σ. This can be written as

$$\left(\frac{h}{mv}\right)^2 \ll \sigma \tag{23}$$

where h is Planck's constant.

The interaction forces between two particles with masses m_1 and m_2 are given by

$$F_1 = m_1 \frac{d^2 r_1}{dt^2} \tag{24}$$

$$F_2 = m_2 \frac{d^2 r_2}{dt^2} \tag{25}$$

where $r_{1 \text{ or } 2}$ is the position vector from the origin of a coordinate system and $F_1 = F_2$. By multiplying (24) and (25) by m_2 and m_1, respectively, and subtracting the two equations, we obtain

$$F_1 = m_r \frac{d^2 (r_1 - r_2)}{dt^2} \tag{26}$$

where the reduced mass m_r is defined as

$$m_r = \frac{m_1 m_2}{m_1 + m_2} \tag{27}$$

Equation (26) states that the collision process is identical to the motion of a particle of reduced mass m_r about a fixed point at a distance $r = r_1 - r_2$ and acted on by one of the forces.

The relative velocity between the two particles can be computed. Since F_1 and r are in the same direction, angular momentum is conserved, so the cross product $F_1 \times r = 0$. This can be written as

$$m_r \frac{d^2 r}{dt^2} \times r = 0 = m_r \left\{ \frac{d[(dr/dt) \times r]}{dt} - \left[\frac{dr}{dt} \times \frac{dr}{dt} \right] \right\}$$

or

$$0 = m_r \left\{ \frac{d[(dr/dt) \times r]}{dt} \right\} \tag{28}$$

since the second term is zero by definition. The conclusion that we can obtain from (28) is that the two particles move in a plane since

$$\frac{dr}{dt} \times r = K \tag{29}$$

Therefore, we need only consider a two-dimensional problem.

In Figure 12.8, the geometry in the plane normal to K of the collision process is shown. The point o is the center of gravity and the lengths of the segments $o1$ and $o2$ are in the ratio of m_2/m_1, and the angle when the two particles are closest

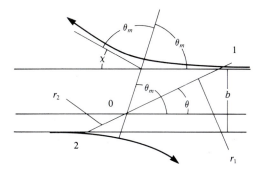

Figure 12.8   Trajectories of particles in a two-dimensional collision.

together is called θ_m. The deflection of particle 1 in the center of mass system is χ, where

$$\chi = \pi - 2\theta_m \tag{30}$$

The distance b is called the *impact parameter*, which is defined as the perpendicular distance between the trajectories when the particles are far from each other.

The total kinetic energy of the particles is given by

$$\mathrm{KE} = \frac{1}{2}m_r\left[\left(\frac{dr}{dt}\right)^2 + \left(r\frac{d\theta}{dt}\right)^2\right] \tag{31}$$

and the total angular momentum about the point o is

$$p_\theta = m_r r^2 \frac{d\theta}{dt} \tag{32}$$

At distances far from the point of interaction,

$$\theta \approx \frac{b}{r}, \qquad \frac{d\theta}{dt} \approx -\frac{b}{r^2}\frac{dr}{dt}$$

and the kinetic energy is found from the expression

$$\mathrm{KE} \approx \frac{1}{2}m_r v_0^2 \tag{33}$$

The angular momentum about the point o is given by

$$p_\theta \approx m_r b v_0 \approx m_r b \frac{r^2}{b}\frac{d\theta}{dt} = m_r r^2 \frac{d\theta}{dt} \tag{34}$$

The velocity v_0 is the relative velocity between the two particles when they are far apart. The total energy U_0 of the system is equal to the sum of the kinetic energy and the potential energy $\phi(r)$, and it is given by

$$U_0 = \frac{1}{2}m_r\left[\left(\frac{dr}{dt}\right)^2 + \left(r\frac{d\theta}{dt}\right)^2\right] + \phi(r) \qquad (35)$$

The Coulomb potential has $\phi(r) \approx -C/r$ for two charged particles. The time variable can be eliminated between (34) and (35), and we obtain the trajectory in polar coordinates as

$$\frac{dr}{d\theta} = \pm\left(\frac{r^2}{b}\right)\sqrt{1 - \left(\frac{b}{r}\right)^2 - \left(\frac{\phi(r)}{U_0}\right)} \qquad (36)$$

At the point of closest approach r_m, which occurs at the angle of closest approach θ_m, we find that

$$1 - \left(\frac{b}{r_m}\right)^2 - \left(\frac{\phi(r_m)}{U_0}\right) = 0 \qquad (37)$$

A solution for (37) does not always exist for all possible charge configurations (dipole, quadrupole, etc.) if $\phi(r_m)$ is an attractive potential, where $\phi(r_m) \approx -\text{constant}/r^n$. For values of $n \geq 2$, there are values of initial energy for which no solution can be found. For a repulsive potential, a solution for (37) can always be found.

The angle of deflection χ can be computed from the integral

$$\chi(b, U_0) = \pi - 2\int_r^\infty \left(\frac{d\theta}{dr}\right)dr \qquad (38)$$

We can substitute (36) into (38) and perform the integration, at least formally.

It is now possible to calculate the differential cross section $\sigma(\chi)$ for the scattering into the angle χ. A particle incident in an annular ring that has an area $2\pi b\,\Delta b$ that is centered on an atom scatters into the solid angle $\Delta\Omega = 2\pi\sin\chi\,\Delta\chi$. Conservation of particles requires that

$$2\pi b\,\Delta b\,Nnv = 2\pi\sin\chi\,\Delta\chi\,Nnv\sigma \qquad (39)$$

where N is the atomic density and n is the density of scattered particles. Hence the collision cross section is written as

$$\sigma(\chi) = \frac{b}{\sin\chi}\frac{\Delta b}{\Delta\chi} \qquad (40)$$

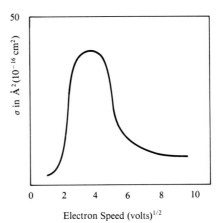

Figure 12.9   Elastic collision cross section of electrons in argon.

The total cross section σ is given by

$$\sigma = \int_0^\pi \sigma(\chi)\, d\Omega \tag{41}$$

This cross section may be infinite for some scattering potentials because $\sigma(\chi)$ is infinite at $\chi = 0$. This states that there is a large cross section for small angle scattering. In Figure 12.9, a representative measured elastic cross-section curve for electrons in argon is presented.

   We could also compute the cross section for momentum transfer in a similar fashion. These are the two elastic collisions of interest in charged particle interaction or collision.

V. Inelastic Collisions

Several different inelastic collisions can be found among the charged-particle interactions and collisions. Each collision has a cross section associated with it. These cross sections for each type of atom or molecule are usually determined from experiment, and graphs appear containing these cross sections as a function of the electric field E divided by the pressure P or as a function of energy of the colliding particle. The reason that the cross section is important is that the number Δn of particles that are undergoing a collision is proportional to the density of target particles N, the density of test particles n, and the thickness of the interaction region Δx. This can be written as

$$\Delta n = \sigma n N\, \Delta x \tag{42}$$

The units of the proportionality constant σ are in terms of an area and hence σ is called the cross section. In the limit of the slab thickness $\Delta x \to 0$, the solution of

(42) yields

$$n = n_0 e^{-N\sigma x} \tag{43}$$

where n_0 is the density of test particles at $x = 0$. In this statement, it has been assumed that the test particles do not disturb the background density of target particles, or $n \ll N$. The background density N is related to the pressure p and temperature T by

$$N = \frac{A}{22,400} \frac{p}{760} \frac{273°}{T°K} \tag{44}$$

where A is Avogadro's number (6.02×10^{23} particles/mole), p and T are normalized by "standard temperature and pressure" (STP) conditions ($T = 0°C = 273°K$ and $p = 760$ mm Hg $= 760$ torr), and there are 22,400 moles in a cubic meter.

Example 12.3   Calculate the mean free path λ_c for collisions at $T = 20°C$ and at a pressure of 10^{-4} torr if the cross section for collision is $\sigma = 10^{-15}$ cm^2.

Answer: The mean free path is defined as the distance where the initial density has undergoine $1/\varepsilon$ collisions. Therefore, $\lambda_c = 1/N\sigma$ and we write this as

$$\lambda_c = \frac{1}{[(A/22,400)(p/760)(273°/T°K)]\sigma}$$

$$= \frac{1}{\{[(6.02 \times 10^{23})/22,400](10^{-4}/760)(273/293)\}10^{-19}}$$

$$= 3 \times 10^6 \text{ meters}$$

We will qualitatively describe several of these inelastic collisions here. The fusion interaction will be described in the next chapter.

Charge Transfer or Charge Exchange

The first type of collision is the transfer of charge from a fast ion to a slow atom. This can be a significant problem if it is desired to confine high-energy ions, since they might collide with a slow neutral particle and escape as a high-energy neutral particle, leaving behind a low-energy ion.

Ionization

If a particle with sufficient energy collides with a neutral particle, it can raise an electron from the valence band to the conduction band or cause it to actually leave the influence of a particular proton and become two separately charged particles: ions and electrons.

Excitation

A particle that collides with another may raise an electron from its present state to another state. From this new state, it may fall back to its original state or to an intermediate state.

Electron Attachment

An electron may attach itself to a neutral atom to form a negative ion. The binding energy is on the order of a few tenths of a volt to a few volts. The attachment energy is greatest for the halogens, since the outer electron shell needs only one electron to be filled.

Recombination

If an electron collides with an ion, the two may recombine to form a neutral particle. An electron could also interact with an ionized molecule, which could experience diassociative recombination:

$$(XY)^+ + e \rightarrow X + Y$$

In either case, excess energy may be emitted as a photon in the first case or as kinetic energy of the fragments in the second case.

Sputtering

If an ion of a gas strikes a surface, it may eject ions from the surface material. These ions may be deposited elsewhere. Sputtering is most prolific if the mass of the gas atom is roughly equal to the surface atom. It is the process that blackens vacuum tubes and lamp bulbs.

VI. Conclusion

From the discussion presented, we note that the motion and confinement of charged particles can be influenced and controlled by electric and and magnetic fields. Charged particles will not cross magnetic field lines that are sufficiently strong and of the proper topology. In many cases this topology is of a simple shape, such as an infinitely long cylinder or a doughnut. The doughnut has an additional problem in that the magnetic field in the "hole" will be different from that at the outer edges, and this inhomogeneity in the magnetic field and the density will create additional particle drifts. Nature has therefore provided many impediments to date in constructing the universal topology that will confine the particles for a long time if their energy and density is sufficiently high. We will examine some of the fundamentals of the state of matter where there are a large number of charged particles distributed in the same region of space in the next chapter, where the subject of a plasma is discussed. A proper treatment of the many collision processes is better left to more advanced discussion than is presented here.

PROBLEMS

1. It is possible to accelerate charged particles to a velocity v, where relativistic corrections have to be included. If this correction becomes $(v/c)^2 \approx 0.1$, find the required accelerating potential ϕ_0 for electrons and argon ions whose atomic weight is 40 times that of hydrogen.

2. Calculate the speed of an electron that has been accelerated through a potential of 1 V. Compare this velocity with a thermal electron that has a thermal velocity due to room temperature, where $v_{th} = (k_B T/m_e)^{1/2}$.

3. Calculate the transverse deflection on a screen of an electron beam that has been accelerated by 300 V if a deflecting voltage of 1 V is applied between two parallel plates. The two parallel plates are separated by 2 cm and are 8 cm wide. The screen is 50 cm from the end of the plates.

4. Find the ratio of the distances (b/a) in a mass spectrometer such that a beam of unknown ions with masses M_a and M_b can be separated from each other and detected.

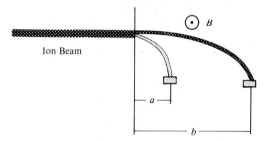

5. Describe the motion of a charged particle that is moving, but is confined within the Van Allen radiation belt depicted in Figure 12.7. Note that the magnetic field is more concentrated at the poles than at the equatorial region. A linear version of this field topology in the fusion community carries the name "mirror machine."

6. The equation of motion for a charged particle in an electric field E can be written as

$$m\frac{\partial v}{\partial t} = qE$$

Describe how this should be modified to include the effects of collisions that occur at the rate of v_C per second.

7. Find the modified equation of continuity that includes the effects of ionization and recombination.

Chapter Thirteen

Plasma Physics

Plasma is the most common state in nature; the solid, the liquid, and the gaseous states comprise less than 1% of the mass of the universe. Even here on earth, we will quickly see the tremendous impact that the ionized state, consisting of almost an equal number of electrons and ions, has on everyday life. The flashing lights of neon signs, the sun that is used to power a solar cell watch, lightning arresters that dissipate the energy carried from a lightning rod to the ground, and the practical plasma processing units that are used to fabricate everything from beer cans to VLSI circuits all have their origins in the plasma state. It is obvious that controlled thermonuclear fusion is the long-term power source of the future. Its raw fuel is the heavy isotopes of hydrogen, which are found in almost an infinite supply in the oceans. Thermonuclear fusion has an uncontrolled relative that can be used for man's eternal quest to create "laboratory" demonstrations of the "big-bang theory." All these involve a detailed understanding of plasma physics. Therefore, we must acquire a modicum of understanding of plasma physics in our early training, as we will certainly see it later in our professional life. Herein we propose to highlight some plasma phenomena and techniques to lead the reader into this fascinating field. Also, the topics should have some importance for the manufacture of the solid-state circuits that were described in the previous pages, and plasma phenomena can be found in many solid-state materials. Waves that exist in gaseous plasmas have their counterpart in solid-state plasmas.

I. Fluid Equations

To describe the behavior of a plasma or any other physical phenomenon, we frequently have to resort to a single equation or to a set of coupled equations. We can invoke various degrees of sophistication to model the plasma, but if we require that "the number of charged particles in a Debye sphere" (this will be explained later) be much greater than 1 and we decide to follow each individual charged-particle motion, we would require approximately 10^{12} coupled three-dimensional nonlinear partial differential equations to describe what is happening in just 1 cubic centimeter of one of the fluorescent lights in a room! Even if we could fathom solving this problem, the answer would not be correct or useful, since quantities such as pressure, temperature, and energy would still have to be defined. Fortunately for us, it is possible to model the plasma with a simpler set of equations. These are based on considering the plasma to be a fluid (i.e., taking some average quantities as being of interest, say a density or a velocity).

The first of these fluid equations are the equations of continuity for electrons,

$$\frac{\partial n_e}{\partial t} + \frac{\partial (n_e v_e)}{\partial x} = 0 \tag{1}$$

and for the ions,

$$\frac{\partial n_i}{\partial t} + \frac{\partial (n_i v_i)}{\partial x} = 0 \tag{2}$$

In these equations, n_e and n_i refer to the electron and ion densities, and v_e and v_i are the electron and ion velocities. These should be vector equations to reflect the three dimensionality of the plasma, but we will simplify our presentation by considering only a one-dimensional situation.

The second group in the set of fluid equations describes the motion of the particles under the influence of various forces, such as the Coulomb forces between the charged particles. Let us assume that no magnetic fields enter into our discussion as this will simplify our work—no vectors are then needed. We write the equation of motion for the electrons as

$$n_e m_e \left(\frac{\partial v_e}{\partial t} + v_e \frac{\partial v_e}{\partial x} \right) = -n_e q E - \frac{\partial P_e}{\partial x} \tag{3}$$

and for the ions as

$$n_i M_i \left(\frac{\partial v_i}{\partial t} + v_i \frac{\partial v_i}{\partial x} \right) = n_i q E - \frac{\partial P_i}{\partial x} \tag{4}$$

In these equations, $q = 1.6 \times 10^{-19}$ coulomb. The masses of the electrons and ions are m_e and M_i, respectively. The second derivative term in both of these equations is a convective term that is required since the local velocities may depend on both position and time; that is,

$$\frac{d}{dt} = \frac{\partial}{\partial t} + \frac{\partial}{\partial x} \frac{dx}{dt} = \frac{\partial}{\partial t} + v \frac{\partial}{\partial x}$$

The terms P_e and P_i are the respective pressures of the electrons and the ions. We could inquire how the pressures evolve in space and time, which would require another group of nonlinear partial differential equations. It is justifiable to close the system at this stage by assuming equations of state for the pressures that are of the form

$$P_e = n_e k_B T_e \tag{5}$$

and

$$P_i = 3n_i k_B T_i \tag{6}$$

The factor of 3 arises because the ions are assumed to be adiabatic with one degree of freedom and the electrons are assumed to be isothermal. Due to their light mass, the thermal conductivity for the electrons is much higher than for the ions, and the electron temperature equilibrates very quickly. The respective temperatures of the electrons and ions are T_e and T_i. Boltzmann's constant has the value $k_B = 1.38 \times 10^{-23}$ joule/kelvin.

We need one more equation to close the system that relates the charged-particle densities to the electric field. This equation is Poisson's equation:

$$\frac{\partial E}{\partial x} = \frac{(n_i - n_e)q}{\varepsilon_0} \tag{7}$$

where $\varepsilon_0 = 1/36\pi \times 10^{-9}$ farads/meter is the permittivity of free space. There are five unknowns, n_e, n_i, v_e, v_i, and E, and five equations, so a solution is theoretically possible. However, the general solution is still too difficult to obtain, so the set must be simplified. This simplification is based on the frequency range of the phenomenon that is being investigated. Electrons can respond to a higher-frequency phenomenon than ions due to their lighter mass, hence their smaller inertia.

II. Low-Frequency Phenomena

The separation of low- and high-frequency effects will be based on the following physical arguments. We initially would like to examine a phenomenon where the velocities of the electrons and the ions are approximately the same. To do this, let $v_e \approx v_i = v$. Therefore, in comparing the magnitudes of the terms in equations (3) and (4) and knowing that $m_e/M_i \ll 1$ ($= 1/1836$ for hydrogen), we can write (3) as

$$0 \approx -n_e qE - \frac{\partial(n_e k_B T_e)}{\partial x} \tag{8}$$

The electric field can be written in terms of a potential ϕ as $E = -\partial\phi/\partial x$. This is called the electrostatic approximation and, as will be shown later, certain low-frequency waves can propagate in this limit and are sometimes given the name *electrostatic waves* to distinguish them from electromagnetic waves, which require the full set of Maxwell's equations. Hence we can solve (8) using

$$0 \approx n_e q\frac{\partial\phi}{\partial x} - \frac{\partial(n_e k_B T_e)}{\partial x} \tag{9}$$

where we will assume that the electron temperature is uniform due to the high thermal conductivity of the electrons. The solution of (9) is

$$n_e = n_0 e^{q\phi/(k_B T_e)} \tag{10}$$

where n_0 is evaluated at a potential $\phi = 0$. These are sometimes called Boltzmann electrons, since we could write

$$\frac{q\phi}{k_B T_e} = -\frac{(m_e v^2)/2}{k_B T_e} = -\left(\frac{v}{v_{th}}\right)^2 \tag{11}$$

where $v_{th} = [2k_B T_e/m_e]^{1/2}$ is the thermal velocity of electrons. With (11) substituted into (10), this becomes a Maxwell–Boltzmann distribution. We must keep in mind that the number of electrons that can reach an increasingly negative potential decreases very quickly to zero.

Let us return to Poisson's equation (7) and write it as

$$\frac{\partial^2 \phi}{\partial x^2} = \frac{(n_e - n_i)q}{\varepsilon_0}$$

$$= \frac{n_0 q [e^{q\phi/(k_B T_e)} - 1]}{\varepsilon_0} \approx \frac{n_0 q \{1 + [q\phi/(k_B T_e)] \cdots - 1\}}{\varepsilon_0}$$

$$\approx \frac{n_0 q^2}{\varepsilon_0 k_B T_e} \phi \tag{12}$$

In writing this, we have made the assumption that the electron and ion densities are almost equal and they have equilibrium value n_0. Also, $|q\phi/k_B T_e| \ll 1$ is assumed.

The solution of (12) is

$$\phi = \phi_a e^{[-x/\lambda_D]} + \phi_b e^{[x/\lambda_D]} \tag{13}$$

where

$$\lambda_D = \sqrt{\frac{k_B T_e \varepsilon_0}{n_0 q^2}}$$

is the electron Debye length. A physical interpretation can be given for this length. Assume that a charge-neutral ($n_e = n_i$) plasma exists initially. What will the potential distribution be if an additional charge is introduced into the plasma? The electrons will redistribute themselves due to Coulomb forces and the electron pressure such that the potential approaches zero within a few Debye lengths (note the analogy to the time constant in an *RC* circuit). This is shown in Figure 13.1.

We can also recall the KdV transmission line described in Chapter 11. This line can serve as an analog for low-frequency plasma phenomena described here.

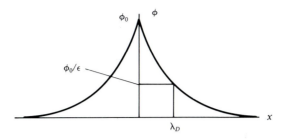

Figure 13.1   Debye shielding of an extra charge at $x = 0$ in a plasma.

The Debye length plays the same role as the section dimension ΔX. Within ΔX, we can only detect microscopic changes, while macroscopic changes occur at the terminals at either end. The tank circuit that appears in the series arm can be thought to be similar to a natural resonance frequency of the ions, the ion plasma frequency. This will be described later.

Hence, from Figure 13.1, we see that the plasma has adjusted itself to shield the single additional charge. This length λ_D has provided a dimension that separates the single charge from the large plasma and is called a Debye *sheath*. Other sheaths will be encountered later.

Example 13.1   A typical laboratory plasma has 10^8 charged particles per cubic centimeter and an electron temperature of $T_e \approx 1$ eV $= 11,600°$K. Compute the Debye length λ_D.

Answer: From

$$\lambda_D = \sqrt{\frac{k_B T_e \varepsilon_0}{n_0 q^2}}$$

$$\lambda_D = \sqrt{\frac{k_B T_e \varepsilon_0}{n_0 q^2}} = \sqrt{\frac{(1.38 \times 10^{-23})(11,600)\{(1/36\pi) \times 10^{-9}\}}{(10^{14})(1.6 \times 10^{-19})^2}} \approx 0.7 \text{ mm}$$

An alternative interpretation is to write λ_D as

$$\lambda_D = \frac{\sqrt{(k_B T_e)/m_e}}{\sqrt{(n_0 q^2)/(m_e \varepsilon_0)}} \tag{14}$$

The numerator has the units of velocity and the denominator must have the units of frequency since λ_D has the units of length. We will call this frequency the electron plasma frequency, which will be discussed in more detail later. Finally, for the potential to actually approach zero, one of the constants of integration ϕ_a or ϕ_b will be set equal to zero in the regions $x < 0$ and $x > 0$, respectively.

Therefore, for a plasma to be a plasma, a common definition is that the number of charged particles in a Debye sphere must be greater than 1; that is,

$$n_0 \frac{4\pi\lambda_D^3}{3} \gg 1 \tag{15}$$

This is sometimes called the plasma parameter. The laboratory plasma described in Example 13.1 has a plasma parameter of approximately 1.4×10^5. For the opposite extreme, we have the single-particle behavior, which was described in the previous chapter, and collective effects will not be important.

III. High-Frequency Phenomena

When we use the term high frequency to describe a phenomenon in a plasma, we usually think of the following physical situation. The frequency is so high that the electrons can easily respond to a force such as an oscillating electric field at this high frequency, but the ions due to their large mass cannot respond. Therefore, the equation of motion for the electrons (3) is important and must be included in a discussion at these frequencies. We will pose the following question: Why is there a "blackout period" during which there is no radio communication with a satellite as it returns to the earth? The answer will tell us much about high-frequency plasma phenomena and will suggest a possible technique for diagnosing a plasma.

Let us assume that an electromagnetic wave with a time dependence given by $e^{i\omega t}$ exists. With this assumption, (3) can be written as

$$(i\omega)n_e m_e \mathbf{v_e} = -n_e q \mathbf{E} \tag{16}$$

In writing (16), we have assumed that the electron pressure term and the convective derivative terms can be neglected. Since the pressure is neglected, this is often called a *cold plasma*. Also, these are vector equations. The definition of the electron current is given by

$$\mathbf{j} = -n_e q \mathbf{v_e} \tag{17}$$

and Maxwell's equations are written as

$$\nabla \times \mathbf{H} = \mathbf{j} + (i\omega)\varepsilon_0 \mathbf{E} \tag{18}$$

and

$$\nabla \times \mathbf{E} = -(i\omega)\mu_0 \mathbf{H} \tag{19}$$

Substitute (16) and (17) into (18), being ever mindful that vector quantities are implied in all equations. We write

$$\nabla \times \mathbf{H} = -n_e q \mathbf{v_e} + i\omega\varepsilon_0 \mathbf{E}$$

$$= -n_e q \left\{ \frac{-n_e q \mathbf{E}}{i\omega n_e m_e} + i\omega\varepsilon_0 \mathbf{E} \right\}$$

$$= i\omega\varepsilon_0 \left\{ 1 + \frac{n_e q^2}{m_e \varepsilon_0 (i\omega)^2} \right\} \mathbf{E}$$

$$= i\omega\varepsilon_0 \left\{ 1 - \frac{\omega_{pe}^2}{\omega^2} \right\} \mathbf{E} \tag{20}$$

The plasma is acting as if it were a relative dielectric constant that could be positive or negative, depending on whether the frequency is above or beneath the electron plasma frequency ($\omega_{pe} = 2\pi f_{pe}$), which is defined as

$$\omega_{pe} = \sqrt{\frac{n_e q^2}{m_e \varepsilon_0}} \tag{21}$$

Using the second of Maxwell's equations (19), it is possible to derive a wave equation:

$$\nabla \times \nabla \times \mathbf{E} = -i\omega\mu_0 \nabla \times \mathbf{H} = -i\omega\mu_0 \left\{ i\omega\varepsilon_0 \left[1 - \frac{\omega_{pe}^2}{\omega^2} \right] \right\} \mathbf{E}$$

or

$$\nabla(\nabla \cdot \mathbf{E}) - \nabla^2 \mathbf{E} = \left(\frac{\omega}{c} \right)^2 \left[1 - \frac{\omega_{pe}^2}{\omega^2} \right] \mathbf{E}$$

This is finally written as

$$\nabla^2 \mathbf{E} + \left(\frac{\omega}{c} \right)^2 \left[1 - \frac{\omega_{pe}^2}{\omega^2} \right] \mathbf{E} = 0 \tag{22}$$

Let us assume that the electric field is polarized in the y direction and depends only on the x coordinate. Hence (22) becomes

$$\frac{\partial^2 E_y}{\partial x^2} + \left(\frac{\omega}{c} \right)^2 \left[1 - \frac{\omega_{pe}^2}{\omega^2} \right] E_y = 0$$

A solution of this wave equation indicates that electromagnetic waves can propagate in the plasma if $\omega > \omega_{pe}$ and they are damped if $\omega < \omega_{pe}$.

In Figure 13.2, a possible microwave diagnostic technique along with a measured frequency response is given. With the electron plasma frequency

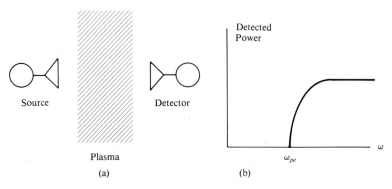

Figure 13.2 (a) Microwave detector of plasma density. (b) Typical detected power as a function of frequency.

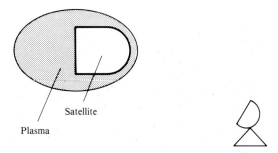

Figure 13.3 Communication with a satellite is interrupted due to plasma layer surrounding it during reentry.

depending on the electron density (21), the cutoff frequency can be used to determine the plasma density. This is frequently used in practice when plasma conditions are such that other diagnostic techniques cannot be applied.

If you have ever listened to a space capsule reentry onto the earth, you know that there is a "blackout" period when there is no radio contact with the capsule. As shown in Figure 13.3, the capsule on its reentry into the atmosphere gets hot due to friction as it passes through the air. It gets so hot in fact that there is enough thermal energy to ionize enough of the air surrounding the capsule such that the local electron plasma frequency is above the frequency of the microwave communication.

Example 13.2   Find the density of a plasma if a microwave at $f = 10$ GHz is cut off, but higher-frequency microwaves can pass through the plasma shown in Figure 13.2.

Answer: At cutoff, the electron plasma frequency $f_{pe} = \omega_{pe}/2\pi$ equals the microwave frequency. Hence

$$f = f_{pe} = \frac{1}{2\pi} \sqrt{\frac{n_e q^2}{m_e \varepsilon_0}}$$

$$10^{10} = \frac{1}{2\pi} \sqrt{\frac{n_e(1.6 \times 10^{-19})^2}{(9.1 \times 10^{-31})[(1/36\pi) \times 10^{-9}]}}$$

from which we compute $n_e = 1.2 \times 10^{19}$ number/m^3 = 1.2×10^{13} number/cm^3.

Before going any farther, let us try to understand what is meant by an electron plasma oscillation. Imagine that we have a slab of uniform density plasma as shown in Figure 13.4 and that a layer of electrons $\Delta n_e = n_e \Delta x$ is displaced to the right. Let us examine the subsequent motion of this electron slab. We will assume that the ions, due to their large mass, do not move and just provide a neutralizing background. At the far left, there will be a bare cloud of ions since the layer of electrons that were there moved to the right to replace those that had moved still farther to the right.

The electron slab obeys the one-dimensional equation of motion (3), which we write as

$$m_e \frac{d^2 \Delta x}{dt^2} = -qE \tag{23}$$

From Gauss's law, the electric field is computed to be

$$E = \frac{\Delta n_e q}{\varepsilon_0} = \frac{n_e q \, \Delta x}{\varepsilon_0} \tag{24}$$

Combine (23) and (24) to yield

$$\frac{d^2 \Delta x}{dt^2} = -\omega_{pe}^2 \, \Delta x \tag{25}$$

which means that the electron slab will just oscillate about its equilibrium value,

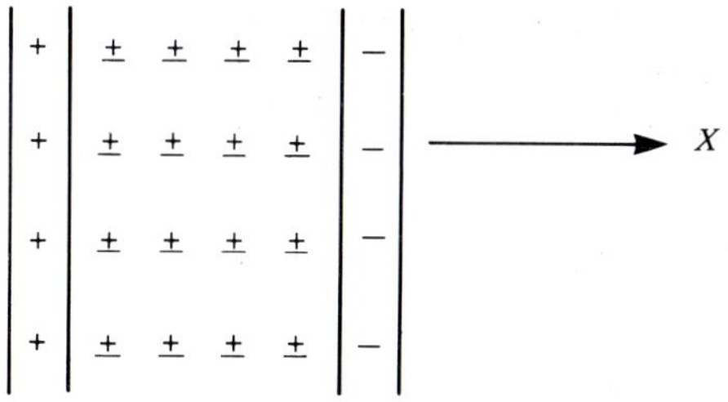

Figure 13.4 Slab model of an electron plasma oscillation.

the frequency of oscillation being the electron plasma frequency

$$\omega_{pe} = \sqrt{\frac{n_e q^2}{m_e \varepsilon_0}}$$

This frequency of oscillation can also be written as

$$f_{pe} = \frac{\omega_{pe}}{2\pi} = \frac{1}{2\pi} \sqrt{\frac{n_e q^2}{m_e \varepsilon_0}}$$

$$= \sqrt{\frac{(1.6 \times 10^{-19})^2}{(9 \times 10^{-31})[(1/36\pi) \times 10^{-9}]}} \frac{\sqrt{n_e}}{2\pi} = 9000\sqrt{n_e} \quad \text{Hz}$$

where n_e is in units of number/cm^3. A similar plasma frequency is found for ions, with the ion mass M_i replacing the electron mass m_e. We might wonder if this oscillation could propagate in a plasma as a wave. It can, as will be described next.

IV. Plasma Waves

There are two plasma waves that can propagate in an unmagnetized plasma. The two waves that are peculiar to a plasma are the low-frequency ion-acoustic wave, which propagates at frequencies below the ion plasma frequency,

$$f_{pi} = \frac{1}{2\pi} \sqrt{\frac{n_0 q^2}{M_i \varepsilon_0}}$$

and the high-frequency electron plasma or Langmuir wave, which propagates above the electron plasma frequency,

$$f_{pe} = \frac{1}{2\pi} \sqrt{\frac{n_0 q^2}{m_e \varepsilon_0}}$$

where $n_e = n_i = n_0$. If a magnetic field were present, the number of possible plasma waves would increase substantially. Finite plasmas will also have their own wave modes, and we can see that the interested student has a lifetime's work ahead before finally closing the book. In this section, we will present the derivation for the ion-acoustic wave since it involves perturbations of both species of the plasma in its characterization. It is also easier to perform experiments at lower frequencies, and the reader may be inclined at some time to do so.

In many plasmas of interest, say a discharge plasma using argon or neon, the temperature of the electrons will be much higher than that of the ions. In fact, we can set $T_e \gg T_i \approx 0$. Hence, (3) and (4) can be combined to yield

$$n_i M_i \left(\frac{\partial v}{\partial t} + v \frac{\partial v}{\partial x} \right) = qE = q \left[-\frac{1}{n_e q} \frac{\partial (n_e k_B T_e)}{\partial x} \right] \tag{26}$$

where we have also included (8). Combine (10) and (26) to obtain (you also could do this directly using $E = -\partial\phi/\partial x$)

$$M_i \left(\frac{\partial v}{\partial t} + v \frac{\partial v}{\partial x} \right) = -k_B T_e \frac{\partial (\ln n_e)}{\partial x} = -q \frac{\partial \phi}{\partial x} \tag{27}$$

Through this reasoning, the set of variables has been reduced to four, n_e, n_i, v, and ϕ, and four equations, (1), (7), (10), and (27), where $E = -\partial\phi/\partial x$ is understood.

To simplify this set still further, we have to introduce the concept of perturbations. The variables will be defined as

$$n_e = n_0 + \Delta n_e$$
$$n_i = n_0 + \Delta n_i$$
$$\phi = 0 + \Delta\phi \tag{28}$$
$$v = 0 + \Delta v$$

In writing this, we have not allowed for any steady drift of the plasma nor any steady-state potential, and the terms with a Δ in front of them are small perturbations about an equilibrium value. Substituting (28) into (2), we write

$$\frac{\partial \Delta n_i}{\partial t} + n_0 \frac{\partial \Delta v}{\partial x} = 0 \tag{29}$$

Substitute (28) into (27), where we have retained only first-order terms, and obtain

$$M_i \frac{\partial \Delta v}{\partial t} = -q \frac{\partial \Delta\phi}{\partial x} \tag{30}$$

Equation (10) can be written as

$$n_0 + \Delta n_e = n_0 e^{(q\Delta\phi)/(k_B T_e)} \approx n_0 \left(1 + \frac{q\Delta\phi}{k_B T_e} \right) \tag{31}$$

and Poisson's equation (7) becomes

$$\frac{\partial^2 \Delta\phi}{\partial x^2} = q \frac{\Delta n_e - \Delta n_i}{\varepsilon_0} \tag{32}$$

Eliminate Δv between (29) and (30):

$$\frac{\partial^2 \Delta n_i}{\partial t^2} - n_0 \frac{q}{M_i} \frac{\partial^2 \Delta \phi}{\partial x^2} = 0 \tag{33}$$

If the electron perturbation Δn_e follows the ion perturbation Δn_i, Poisson's equation (32) will be simplified. Therefore, let $\Delta n_e = \Delta n_i$ in (31) and substitute it in (33):

$$\frac{\partial^2 \{n_0 [(q \Delta \phi)/(k_B T_e)]\}}{\partial t^2} - n_0 \frac{q}{M_i} \frac{\partial^2 \Delta \phi}{\partial x^2} = 0 \tag{34}$$

or

$$\frac{\partial^2 \Delta \phi}{\partial t^2} - \left[\frac{k_B T_e}{M_i} \right] \frac{\partial^2 \Delta \phi}{\partial x^2} = 0 \tag{35}$$

The term

$$\sqrt{\frac{k_B T_e}{M_i}}$$

has the units of velocity, and this velocity is called the ion-acoustic velocity, which we will give the symbol c_S. For an argon plasma (atomic mass = 40 × mass of hydrogen) with an electron temperature $T_e = 1$ eV, this velocity is $c_S \approx 1.5 \times 10^3$ m/s. Note that (35) is a linear wave equation of the type that was examined in Chapter 1. We would therefore expect that the plasma would support the propagation of various waves, one of them being the ion-acoustic wave.

This wave is very similar to the propagation of ordinary acoustic waves in a gas. In that case there is a compression and a rarefaction of the density of gas molecules as the wave propagates. Here there is a similar compression and rarefaction of the density of charged particles. These perturbations in the density are coupled together by the local electric field that is set up by the density perturbations, and the perturbation propagates with a velocity equal to c_S. The density perturbations in either an ordinary sound wave or in this ion-acoustic wave are in the direction that the wave is propagating. These are called *longitudinal waves*. These waves should be contrasted with waves such as electromagnetic waves or waves on a string where the perturbation is transverse to the direction of propagation, which are called *transverse waves*.

Let us now remove the restriction that $\Delta n_e \approx \Delta n_i$, which removed Poisson's equation from our consideration. The procedure of deriving a wave equation will be the same as before although a few more terms will appear. Write Poisson's equation (32) as

$$\frac{\partial^4 \Delta \phi}{\partial x^2 \, \partial t^2} = \frac{\left(\dfrac{\partial^2 \Delta n_e}{\partial t^2} - \dfrac{\partial^2 \Delta n_i}{\partial t^2} \right) q}{\varepsilon_0} = \left\{ \left(\frac{n_0 q}{k_B T_e} \right) \frac{\partial^2 \Delta \phi}{\partial t^2} - \left(\frac{n_0 q}{M_i} \right) \frac{\partial^2 \Delta \phi}{\partial x^2} \right\} \frac{q}{\varepsilon_0}$$

Let

$$\Omega_{pi} = \sqrt{\frac{n_0 q^2}{M_i \varepsilon_0}}$$

be the ion plasma frequency in radians/second. Note from (14) that the ratio of the ion-acoustic velocity to the ion plasma frequency is equal to the Debye length. This equation can be written in a simpler form as

$$\frac{\partial^4 \Delta\phi}{\partial x^2 \partial t^2} = \left(\frac{\Omega_{pi}}{c_s}\right)^2 \frac{\partial^2 \Delta\phi}{\partial t^2} - \Omega_{pi}^2 \frac{\partial^2 \Delta\phi}{\partial x^2} \tag{36}$$

It is possible to obtain a wave solution for (35) or (36) of the form

$$\Delta\phi = \Delta\phi_0 e^{i(kx - \omega t)} \tag{37}$$

where $i = (-1)^{1/2}$, ω is the angular frequency ($\omega = 2\pi f$), and k is the wave number ($k = 2\pi/\lambda$, where λ is the wavelength). Substitute (37) in (36) and obtain (the term $\Delta\phi$ factors out after the differentiation)

$$(ik)^2(-i\omega)^2 = -\left(\frac{\Omega_{pi}}{c_s}\right)^2 (-i\omega)^2 - \Omega_{pi}^2 (ik)^2$$

or

$$\left(\frac{\omega}{k}\right)^2 = \frac{c_s^2}{1 + (kc_s/\Omega_{pi})^2} \tag{38}$$

This can also be written as

$$1 + \frac{(\Omega_{pi}/c_s)^2}{k^2} - \left(\frac{\Omega_{pi}}{\omega}\right)^2 = 0 \tag{39}$$

A sketch of (39), which is the dispersion relation for an ion-acoustic wave, is shown in Figure 13.5. It is plotted in two different ways. Recall that the phase velocity v_ϕ of a wave is defined as $v_\phi = \omega/k$. The fact that the propagation velocity

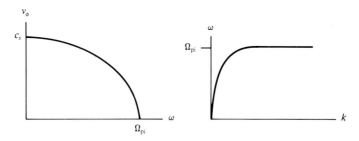

Figure 13.5   Dispersion relation for ion-acoustic waves.

depends on frequency implies that pulses that contain many frequencies (i.e., many Fourier components) will be distorted as it propagates. The high-frequency terms will trail behind the lower-frequency terms. The low-frequency phase velocity is just $c_s = [k_B T_e/M_i]^{1/2}$, which is given in (35).

The ratio $(kc_S/\Omega_{pi})^2 = 1$ in (38) can also be used to understand the physical mechanism causing this dispersion. The ration c_S/Ω_{pi} is equal to the Debye length λ_D (recall that the mass appears in both of the terms c_S and Ω_{pi}). Therefore, when the wavelength λ of the ion-acoustic wave decreases and becomes on the order of the Debye length, the plasma cannot respond and charge separation ensues between the particles.

By starting with the fluid equations as a description for the plasma and with the assumption of $T_e \gg T_i \approx 0$, at this stage we are missing out on one fundamental feature of plasma waves, the damping of the waves due to the interaction of the waves with the plasma. It turns out that the longitudinal electric field of the wave will tend to accelerate or decelerate the plasma particles if the wave velocity defined by (36) is on the order of the ion thermal velocity (i.e., approximately $[3k_B T_i/M_i]^{1/2}$. If we use a kinetic theory to model the plasma and think of the distribution function as a collection of beams with different velocities [say the Maxwell–Boltzmann distribution given in (10) with the thermal velocity of ions replacing the thermal velocity of electrons], the plasma particles with a velocity about equal to the phase velocity of the wave will interact with the wave. See Figure 13.6. Particles that are slowed down will give energy to the wave, and those initially slower than the wave will be sped up and extract energy from the wave. If more wave energy is lost than gained, the wave will damp. This is called Landau damping in honor of the theoretical physics giant who correctly worked out the mathematics.

It has several practical applications. First, energy must be conserved, so the energy that is extracted from the wave and given to the plasma will increase the random motion in the plasma and thus heat it. This is important for fusion applications. Second, if the distribution function has the opposite slope as the velocity of the wave, energy will be extracted from the plasma and given to the wave, which will cause it to amplify. This leads to a practical high-frequency microwave tube, the traveling-wave amplifier described in the next chapter.

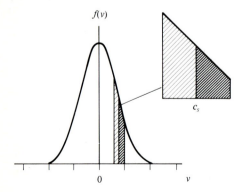

Figure 13.6 Distribution function $f(v)$ as a function of velocity v. Inset shows particles about the wave velocity.

Example 13.3   Describe the effects on the propagation of an ion-acoustic wave in argon if a small amount of hydrogen leaks into the plasma chamber. Assume the temperature ratio of electrons to ions is $T_e/T_i \approx 40/3$.

Answer: The velocity of the ion-acoustic wave is given by $c_S = [k_B T_e/M_{Ar}]^{1/2}$ and the thermal velocity of the hydrogen ions

$$v_{th} = \sqrt{\frac{3k_B T_i}{M_H}} = \sqrt{\frac{3T_i}{T_e}} \sqrt{\frac{M_{Ar}}{M_H}} \sqrt{\frac{k_B T_e}{M_{Ar}}}$$

Hence these thermal ions are moving with approximately the ion-acoustic velocity and Landau damp the wave. This particular demonstration of Landau damping can be easily shown in the laboratory and can arise if oil vapor from a vacuum pump escapes into the chamber.

The derivation for the electron plasma wave or the Langmuir wave is similar to the one we have just carried out for the ion-acoustic wave. It is actually easier, since at the high frequencies above the electron plasma frequency, the ions can be considered to form a neutralizing stationary background. We need to consider the equations of continuity (1) and motion (3) for the electrons, an equation of state (5), and Poisson's equation (7). Following the procedure outlined previously, we can obtain a dispersion relation for this wave of the form

$$\omega^2 = \omega_{pe}^2 + 3k^2\left(\frac{k_B T_e}{m_e}\right)$$

This Bohm–Gross dispersion relation is similar to the dispersion relation of an ordinary electromagnetic wave propagating in waveguide in that there is a definite low-frequency cutoff and that it approaches the electron thermal velocity $[k_B T_e/m_e]^{1/2}$ at short wavelengths. We would expect and do indeed find that this wave interacts with these thermal electrons and is heavily Landau damped.

V. Plasma Production: Richardson–Dushman Equation

Prior to further studying about waves, we must first inquire how we make a plasma, since this technique will also be encountered when we use any vacuum tube. We also must ask how to excite a wave and how to detect a wave. Fortunately, to create a quiescent plasma, all we have to do is have some source of energy that is located apart from the region where the experiments are to be performed and that is greater than the ionization potential of the neutral atom. We could use heat or light, but the required intensities would be too high for easy laboratory operation. Boiling electrons from joule heated filaments and then accelerating them to energies of 50 eV is relatively easy. If these electrons collide with a neutral atom, their energy will be sufficient to ionize the atom, whose ionization potential say for argon is approximately 15 eV.

A fundamental equation in studying plasma or tube phenomena is the Richardson–Dushman equation. It allows us to compute the current density that can be emitted from a thermally heated surface. That is, how many electrons or what is the current that we can obtain from a hot cathode? Once the electrons leave a hot surface, they can be used either to make plasmas or in vacuum tubes such as cathode-ray tubes. Several methods can be used to obtain this equation. We will outline one here for thermodynamics by considering a sea of electrons adjacent to the hot metal surface (cathode) as an *ideal gas*, as shown in Figure 13.7. (An alternative derivation using the elements of quantum mechanics is given in Appendix B). By assuming that the electrons act as an ideal gas of *noninteracting* particles, we can use the laws of thermodynamics to describe their behavior. In particular, the best efficiency that we can ever obtain in a thermodynamic engine is the *Carnot efficiency*. We define this efficiency η as

$$\eta = \frac{\text{work out}}{\text{heat in}} \tag{40}$$

where

Work out = force $\times$ displacement

$$= F\,\Delta L = \left(\frac{F}{A}\right) \times (A\,\Delta L) = P \times \Delta(AL) = P\,\Delta V$$

$$= \text{pressure} \times \text{volume change} \tag{41}$$

If the temperature of the electron gas stays the same, we can write this as

$$\Delta(PV) = P\,\Delta V + V\,\Delta P = \mathscr{R}\,\Delta T \approx 0 \tag{42}$$

where we have made use of the ideal gas law: $PV = \mathscr{R}T$ and $\mathscr{R} = 8.31 \times 10^3$ (joules)/(kg-mole-°K). Carnot efficiency is also given by

$$\eta = \frac{\Delta T}{T} \tag{43}$$

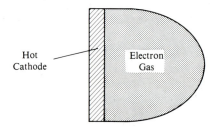

Hot
Cathode

Electron
Gas

Figure 13.7   Cloud of electrons emitted from the cathode is considered to behave as an ideal gas.

Therefore,

$$\frac{\Delta T}{T} = \frac{V \Delta P}{Q} \tag{44}$$

where Q is the heat input. Solving for Q, we write

$$Q = VT \frac{\Delta P}{\Delta T}$$

Eliminate V from this equation using the ideal gas law $V = \mathcal{R}T/P$ and write

$$Q = \frac{\mathcal{R}T}{P} T \frac{\Delta P}{\Delta T}$$

or

$$\frac{\Delta P}{P} = \frac{Q}{\mathcal{R}T^2} \Delta T \tag{45}$$

For an ideal gas, we can relate the heat input to temperature by

$$Q = \frac{5}{2} \mathcal{R}T + Q^* \tag{46}$$

where Q^* is a constant. Therefore,

$$\frac{\Delta P}{P} = \frac{5}{2} \frac{\Delta T}{T} + \frac{Q^*}{\mathcal{R}} \frac{\Delta T}{T^2} \tag{47}$$

This can be integrated to yield

$$P = T^{5/2} e^{-Q^*/\mathcal{R}T} \tag{48}$$

The current density is given by

$$j = -nq\langle v \rangle$$

where n is given by an isothermal equation of state:

$$n = \frac{P}{k_B T}$$

The average thermal velocity $\langle v \rangle$ of the electrons is given by

$$\langle v \rangle = \sqrt{\frac{k_B T}{m_e}}$$

The current is finally written by combining these equations to yield

$$j = A'T^2 \exp\left(-\frac{\text{work function}}{k_B T}\right) \qquad (49)$$

where $A' = 1.2 \times 10^6 \, \text{A}/(\text{m-}°\text{K})^2$ and it is a constant! It has to be a constant since the derivation followed from the basic laws of thermodynamics. If you do an experiment and measure a different value, either you are wrong or else thermodynamics is wrong! Guess which answer you should think is correct! This is the Richardson–Dushman equation for calculating the current emitted from a thermally heated filament or cathode.

A typical value of energy that is required to cause electron emission is approximately 1 to 4 eV. We can make a coating to put on the metal to increase the emission. For example, a mixture of

$$\text{SrCO}_3 + \text{BaCO}_3 \longrightarrow \text{SrO} + \text{BaO} + 2\text{CO}_2$$

when sprayed in an alcohol mixture on nickel works fairly well. The carbon dioxide is liberated first when heated and electrons are emitted at $T \approx 1100°\text{K}$.

Electrons are now available from a thermally heated cathode and can be accelerated to either ionize a gas or be used in a cathode-ray tube. The next question that must be examined is whether there is an upper limit to this current.

VI. Child–Langmuir Space-Charge-Limited Current

The Richardson–Dushman equation allows us to predict the current that can be emitted from a cathode. Let us ask, what effect will there be on the emitted current if some of the electrons remain in the vicinity of the cathode? Will there be a reduction in this current due to these recalcitrant electrons, which we will call *space charge*? The answer is, of course, that there will be an effect. We will derive it here.

Consider the charge distribution shown in Figure 13.8. Some electrons emitted from the cathode remain in the vicinity of the cathode. They will tend to prevent other electrons from leaving the cathode, hence reducing the current. We can estimate this reduction as follows. First write Poisson's equation:

$$\frac{d^2\phi}{dx^2} = \frac{n_e(x)q}{\varepsilon_0} \qquad (50)$$

where the electron density n_e is a function of position (this is as yet an unknown function). At $x = 0$, ϕ and $d\phi/dx$ are chosen to be both equal to zero since the cathode is a metallic structure, and it is grounded.

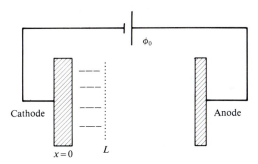

Figure 13.8   Space charge in front of a cathode.

To calculate $n_e(x)$, look at the steady-state time-independent equation of continuity for the electrons (1). The time dependence can be neglected since it is difficult to heat a metal object very quickly. Hence we write

$$\frac{d(n_e v_e)}{dx} \approx 0 \tag{51}$$

whose solution is

$$n_e v_e = NV \tag{52}$$

where NV is a constant of integration. Therefore, it is possible to write the electron density distribution in terms of the distribution of the electron velocity. If an electron falls through a potential difference ϕ, it will gain a kinetic energy equal to that of the potential difference.

$$q\phi = \frac{m_e v_e^2}{2} \tag{53}$$

or

$$v_e = \sqrt{\frac{2q\phi}{m_e}} \tag{54}$$

Therefore, Poisson's equation becomes

$$\frac{d^2\phi}{dx^2} = \frac{NV}{v_e}\frac{q}{\varepsilon_0} = \frac{NV}{\sqrt{2q\phi/m_e}}\frac{q}{\varepsilon_0} \tag{55}$$

This can be integrated since only one dependent variable, ϕ, appears. The first integral is

$$\frac{1}{2}\left(\frac{d\phi}{dx}\right)^2 = 2K\sqrt{\phi} + A \tag{56}$$

where

$$K = \frac{NVq}{\varepsilon_0\sqrt{2q/m_e}}$$

and A is a constant of integration. We set $A = 0$ due to the boundary conditions at $x = 0$ and write

$$\phi^{-1/4}\,d\phi = \sqrt{4K}\,dx$$

The second integral of (55) is finally

$$\frac{\phi_0^{3/4}}{3/4} = \sqrt{4K}\,L + B \tag{57}$$

where again the constant of integration $B = 0$ due to boundary conditions. The dimension of the space-charge region is L, and the potential on the anode is ϕ_0. The potential distribution is now known in the space-charge region.

$$\phi = \left[\frac{3}{4}\sqrt{4K}\,x\right]^{4/3} \tag{58}$$

Eliminate the constant K to finally obtain

$$\phi = \left[\frac{3}{4}\sqrt{4\frac{NVq}{\varepsilon_0\sqrt{2q/m_e}}}\,L\right]^{4/3}$$

The current from the cathode is defined as

$$j = -NVq \tag{59}$$

or

$$j = \frac{4\varepsilon_0/9\sqrt{2q/m_e}(\phi_0)^{3/2}}{L^2}$$

This is the Child–Langmuir space-charge-limited current relation. This is the maximum current that can be emitted from a cathode if space charge is included. To improve the current-handling capacity, methods must be found to neutralize this charge or "sweep it away." In the history of tube development, considerable effort has been expended to do just this.

Both the Richard–Dushmann and the Child–Langmuir equations, which have been described in the previous section and in this section, are important for the development of laboratory devices in which to carry out plasma experiments. This will be noted in the next section, where laboratory plasmas are considered.

VII. Laboratory Plasmas

Since we want a "clean" plasma environment in which to do experiments (i.e., not too many unwanted particles around to mess up the experiment), the experiment has to be done under vacuum conditions where the numbers of free neutrals are few and far between. Space is such a nice environment and it almost satisfies one of the theorist's premises of an infinite, homogeneous plasma. All that is needed to perform a laboratory experiment are large vacuum pumps and vacuum chambers, say on the order of $\frac{1}{2}$ cubic meter in size. The pressure should be less than 10^{-6} torr and a known gas such as argon is bled into the vessel until the pressure is 10^{-4} torr (atmospheric pressure is 760 torr or 760 mm Hg). Then any impurities will be less than 1% of the plasma environment. Continuous pumping will also be employed to keep this level of purity at all times, barring an unforeseen accident.

A useful plasma environment can be created using the magnetic cage shown in Figure 13.9, which consists of approximately 1000 to 1500 small permanent magnets inserted into a square aluminum channel that completely encloses the plasma region. The magnets have their pole faces in the same direction and each row has alternating pole faces. Why magnets? Recall the discussion in Chapter 12 on charged-particle ballistics, where we looked only at one-dimensional situations. This magnetic cage confines the "ionizing" electrons emitted from the filaments and accelerated with a dc voltage supply connected to the cage structure. This voltage is greater than the ionization potential of the background gas. These electrons are accelerated to the cage structure, which serves as the anode, and stay within the cage until they have time to ionize the neutral atom. Remember that the electrons will experience a force that is perpendicular to both their velocity and the direction of the magnetic field. The electrons experience a force that is given by

$$\mathbf{F} = q(\mathbf{v} \times \mathbf{B}) \tag{60}$$

where the bold type indicates a vector. If we set this force equal to the centrifugal force

$$F = \frac{mv^2}{r} \tag{61}$$

where we will keep track of the vector directions in our minds, we can solve for the Larmor radius for the electrons by equating (60) and (61). We find that

$$r_L = \frac{mv^2}{qvB} = \frac{mv}{qB} \tag{62}$$

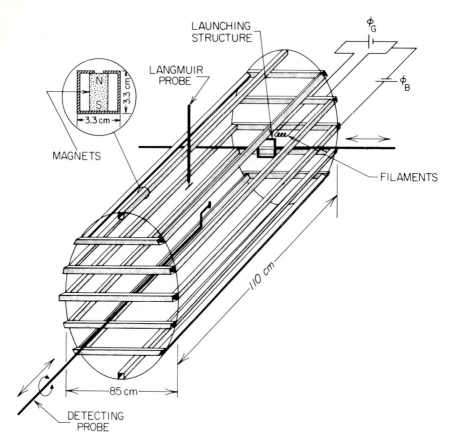

Figure 13.9   Typical magnetic-cage structure in which wave experiments can be performed. The entire structure is inserted in a vacuum chamber.

It turns out that this radius for the ionizing electrons is very small. You should put in some numbers for a B field of 100 gauss (1 tesla $= 10^4$ gauss). The plasma environment that has been created in the laboratory has the following typical numbers; there are approximately 10^8 electrons and ions in every cubic centimeter and the plasma has a temperature of approximately 1 eV (1 eV $=$ 11,600 K). Actually, there may be a few more ions than electrons in the laboratory plasma. Due to their greater mobility, some of the electrons may escape along a B field line, causing the plasma to have a slightly positive potential. The potential of the plasma is called the plasma potential.

The next question is concerned with how we measure these numbers in the laboratory and the effect of introducing perturbing objects that may be biased and/or conducting into the plasma. The plasma will adjust itself to these foreign objects and shield itself by enclosing the object in a sheath.

VIII. Plasma Sheaths

If we insert a metal object in a plasma, the plasma particles will adjust their spatial distribution so that the effect of the object will be *screened out* and any potential that is applied to it will not be detected at large distances from the object. Your first encounter with such a screening occurred when you examined the details of the *Debye length* or *Debye sheath*, where the effects of an additional charge inserted in a large plasma rapidly decreased as we moved away from it. An analogous "screening" occurs when a voltage step is applied to an *RC* circuit and its value decreases in a "time constant." Two sheaths will be described, a steady-state Langmuir sheath and a transient sheath, which was discovered in experiments on ion-acoustic waves. The transient sheath has a solid-state analog to the depletion layer that is found in *pn* junction diodes.

Before describing a steady-state sheath, we have to bring in the idea of diffusion, in particular ambipolar diffusion. If an enhanced concentration of plasma existed in some localized region, we might expect that the local concentration would decrease to the average density and the excess charged particles would spread over all the region. If a glass of milk tips over on a table, the milk quickly covers the entire table (i.e., it diffuses). The same occurs for a plasma, but the additional feature peculiar to a plasma is that the individual particles have a charge of $+$ or $-$, so electrical forces will also influence the particles' motion.

The flux of electrons is given by

$$j_e = -D_e \frac{\partial n_e}{\partial x} - n_e \mu_e E \tag{63}$$

and for ions the flux is given by

$$j_i = -D_i \frac{\partial n_i}{\partial x} + n_i \mu_i E \tag{64}$$

For simplicity, we will assume a one-dimensional system, so we do not need vectors. The symbol $D_{e\,or\,i}$ is the diffusion coefficient, and $\mu_{e\,or\,i}$ is the mobility of the electrons or ions, respectively. The diffusion and mobility constants are related through the Einstein relations

$$\frac{D}{\mu} = \frac{k_B T}{q} \tag{65}$$

Since both species of particles move together, $j_e \approx j_i$. Therefore, solve for E between (63) and (64):

$$E = \frac{-D_e(\partial n_e/\partial x) + D_i(\partial n_i/\partial x)}{n_i \mu_i + n_e \mu_e} \tag{66}$$

Since the particles diffuse together, we can write

$$\frac{\partial(n_i/n_e)}{\partial x} = 0 = \frac{n_e(\partial n_i/\partial x) - n_i(\partial n_e/\partial x)}{n_e^2} \tag{67}$$

or

$$\frac{\partial n_i}{\partial x} = \frac{n_i}{n_e}\frac{\partial n_e}{\partial x} \tag{68}$$

Substituting (68) and (66) in (63), we write

$$j_e = -\frac{n_i\mu_i D_e + n_i\mu_e D_i}{n_i\mu_i + n_e\mu_e}\frac{\partial n_e}{\partial x} = -D_A\frac{\partial n_e}{\partial x} \tag{69}$$

where D_A is the ambipolar diffusion coefficient. Employing the plasma condition that the electron and ion densities are almost equal and with the electron temperature being greater than the ion temperature, (69) simplifies to

$$j_e \approx -\frac{\mu_i\mu_e}{\mu_i + \mu_e}\frac{k_B T_e}{q}\frac{\partial n_e}{\partial x} \tag{70}$$

With the Boltzmann assumption for electrons, we write

$$\frac{\partial n_e}{\partial x} = n_e\frac{q}{k_B T_e}\frac{\partial \phi}{\partial x} = -n_e\frac{q}{k_B T_e}E$$

where we recall that $E = -\partial\phi/\partial x$.

The mobility of the charged particles is calculated in the same manner as it was calculated for solids previously. Consider the following situation. Charged particles exist in a region of space in which an electric field exists. These particles will be accelerated due to the electric field until they collide with another particle, say a neutral one. This collision will occur (on the average) after a collision time τ_{coll}. If the particle started out initially with zero velocity, its final velocity can be computed from

$$m\frac{dv}{dt} = qE$$

yielding a final velocity of

$$v_{final} = \frac{q\tau_{coll}}{m}E$$

This holds true for both ions and electrons. The average velocity is

$$v_{ave} = \frac{v_{final} + v_{initial}}{2} = \frac{q\tau_{coll}}{2m} E$$

or

$$v_{ave} = \mu E$$

Hence we define the mobility $\mu = q\tau_{coll}/2m$, where the appropriate value of mass is chosen for the electrons or ions. Note that this model has assumed that the particle starts with zero velocity *after each collision*. More complicated models do exist and they yield a different numerical coefficient.) Therefore,

$$\frac{\mu_e \mu_i}{\mu_e + \mu_i} \approx \mu_i = \frac{q\tau_{coll}}{2M_i}$$

and (70) can be written as

$$j_e = -\frac{\tau_{coll}}{2} \frac{k_B T_e}{M_i} \frac{\partial n_e}{\partial x} \tag{71}$$

Note the appearance of the ion-acoustic velocity

$$c_s = \sqrt{\frac{k_B T_e}{M_i}}$$

in this derivation. Hence the ambipolar diffusion coefficient can be written as

$$D_A = \frac{c_s^2 \tau_{coll}}{2} \tag{72}$$

We will let the ambipolar diffusion current be given by

$$j_{ambipolar} = nqc_S \tag{73}$$

since we have seen that the ion-acoustic velocity appears when the two species drift together. The ions provide the inertia due to their heavier mass, and the electrons provide the pressure since we have assumed that the electron temperature is greater than the ion temperature.

An interesting question arises if this ambipolar diffusion current is set equal to the space-charge-limited current. Both are steady-state currents. Therefore, let

$$\frac{(4\varepsilon_0/9)\sqrt{\dfrac{2q}{m_e}}\,\phi_0^{3/2}}{L^2} = nqc_s \tag{74}$$

In a given plasma with known density and temperature, a space-charge sheath is set up that extends a distance L into the plasma from a metal object that depends on the voltage bias applied to the object. This is called a Langmuir sheath, and its size is computed from (74). Its size has a nonlinear voltage dependence of $L \approx \phi_0^{3/4}$.

The second sheath that we will study here is a transient sheath. Imagine the following experimental situation. A metal plate is inserted in a plasma and it is connected through a switch to a battery, as shown in Figure 13.10. At $t = 0$, the switch is closed and, due to the polarity of the battery, the electrons are instantly repelled from the vicinity of the plate. The ions will not initially move due to the difference in mass, but the electrons can be easily blown away to at least some distance W. The problem that remains is to compute this distance W.

Only ions remain in the region $0 \le x \le W$, so we need only solve Poisson's equation in this region, which is given by

$$\frac{d^2\phi}{dx^2} = -\frac{n_i q}{\varepsilon_0} \tag{75}$$

If the ion density in front of the plate is homogeneous, the solution is

$$\phi = -\frac{n_i q}{\varepsilon_0}\frac{x^2}{2} + Ax + B \tag{76}$$

The boundary conditions are that at $x = 0$

$$\phi = -\phi_0$$

and at $x = W$

$$\phi = 0 \quad \text{and} \quad \frac{d\phi}{dx} = 0$$

Here we have neglected any steady-state Langmuir sheath that may be present due to another dc voltage bias. Employing these three boundary conditions in

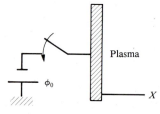

Plasma

Figure 13.10   Switch connecting the battery to the electrode is closed at $t = 0$.

(76), we find that

$$W = \sqrt{\frac{2q\phi_0}{n_i q^2 / \varepsilon_0}} \tag{77}$$

The width of this sheath, which is called a transient sheath, is similar to the depletion width in a *pn* junction and, upon reflection, the assumptions made in its derivation are similar to those made in deriving the depletion width.

Writing (77) in this format allows the reader to make the analogy to the Debye sheath by just replacing battery energy $2q\phi_0$ by the thermal energy $k_B T_e$. As time increases beyond the initial change of voltage, which occurs in a very short time span that is much less than the ion plasma period, the ions will start to move, and we have to solve the full set of nonlinear coupled partial differential equations presented earlier in order to predict the plasma response as it approaches a new steady-state condition.

IX. Langmuir Probe

There are several techniques to measure the parameters of a plasma such as the average number density of electrons or the temperature of the electrons. We can use the cutoff frequency $f = f_{pe}$ in a microwave diagnostic technique to determine an electron number density, an energy analyzer to determine an ion temperature, and in many other sophisticated devices and techniques involving lasers to determine temperatures and densities. Herein we will examine just one tool, the Langmuir probe. This probe is named in honor of the father of plasma physics, Irving Langmuir. He was a scientist at General Electric in the 1920s and was studying gas discharges, when he noted some peculiar oscillations on a detector. He said that they oscillate like a plasma and this name has remained in use. (Fortunately, he didn't have some Jello in front of him at the time, since certain scientists today would not like to be called Jello physicists.) The probe that will be described is still a fundamental tool in plasma physics and an understanding of it will summarize many plasma properties. Langmuir was later awarded the Nobel prize for his work in chemistry.

Consider the experimental setup shown in Figure 13.11. A small metal probe is inserted into the plasma. Both the current collected by the probe and the voltage between the probe and ground are monitored, and the indicated idealized current–voltage characteristics are measured. Remember that some of the higher-mobility electrons have left the plasma, so the plasma potential ϕ_{pl} is positive. The plasma potential is the potential that the plasma naturally assumes in an equilibrium situation between plasma production and plasma losses through recombination or in losses to the wall. We find by experiment that the "cleanliness" of a plasma experiment can be determined by monitoring this

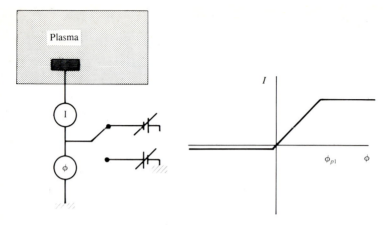

Figure 13.11   Langmuir probe circuit and measured characteristics.

plasma potential. A "dirty" chamber will not permit the escape of excess electrons, and the plasma potential will be negative.

Let us now try to interpret the curve shown in Figure 13.11. For voltages that are very positive ($\phi \gg \phi_{pl}$), the current will be due to the electrons drawn from the plasma. It will be almost a constant due to the steady-state sheath that surrounds the metallic probe, and it is given by

$$I_{\text{very positive}} = n_e q v_{the} A \tag{78}$$

where A is the area of the probe and v_{the} is the velocity of the electrons to the probe. It can be shown to be equal to

$$v_{the} = \sqrt{\frac{k_B T_e}{m_e}}$$

If the probe is very negative, the current will be due to ions, and it is given by

$$I_{\text{very negative}} = n_i q c_s A \tag{79}$$

where

$$c_s = \sqrt{\frac{k_B T_e}{M_i}}$$

when $T_e \gg T_i$. The electrons will be dragged along by the ions and they will drift together.

Remember that we have assumed that the electrons in this plasma are Boltzmann electrons. Due to the mass difference between the currents given in (78) and (79), the majority of the current even for potentials that are slightly negative ($\phi \gtrsim \phi_{pl}$) will still be due to the electrons. In fact, it will given by

$$I_{\text{slightly negative}} \approx n_e q v_{the} A e^{q\phi/(k_B T_e)} \tag{80}$$

Now if we measure the current at two values of ϕ, say $\phi = \phi_{pl}$ and a value such that it is ε^{-1}, we can compute T_e from $q\phi/k_B T_e = -1$. You could also plot the values of current versus voltage on semilog graph paper and examine the slope of the linear region. Now knowing the temperature of electrons from (80) and the cross-sectional area of the probe, the density can be computed from (78) from the measured current at $\phi \gg \phi_{pl}$. This is one method to measure the plasma parameters and is probably the standard and most widely employed technique among the plasma physics aficionados.

Example 13.4   Using a Langmuir probe, we measure the following voltage–current curve. The area A of the probe was 5 mm^2 = 5×10^{-6} m^2. Determine the plasma parameters ϕ_{pl}, T_e, and n_e.

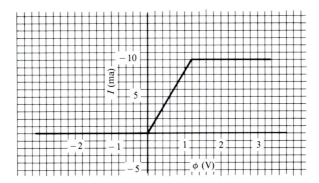

Answer: The knee of the Langmuir curve occurs at $I = 10$ mA and $\phi = +1.2$ V. The plasma potential ϕ_{pl} is therefore at $+1.2$ V. The electron temperature T_e is computed from the curve in the range $0 \leq \phi \leq 1.2$ V, where only the Maxwellian electrons contribute and the ion flow can be neglected. Using (80), we note that the current decreases to $(10 \text{ mA})(\varepsilon^{-1}) \approx 2.6$ mA at ≈ 0.3 V. Therefore, the electron temperature $T_e \approx (1.2 - 0.3)$ eV $= 0.9$ eV. The electron density n_e is computed by (78) as

$$n_e = \frac{I_{\text{very positive}}}{q v_{the} A}$$

where

$$v_{the} = \sqrt{\frac{k_B T_e}{m_e}}$$

$$= \sqrt{\frac{(1.38 \times 10^{-23})(0.9 \times 11,600)}{9.1 \times 10^{-31}}} = 4 \times 10^5 \text{ m/s}$$

Therefore, $n_e \approx 3.4 \times 10^{16}/\text{m}^3 = 3.4 \times 10^{10}/\text{cm}^3$.

X. Wave Experiments

Having created a plasma that is quiescent, large, and relatively homogeneous, we are now going to study what can be done with it in the laboratory. We will focus this discussion on the ion-acoustic wave since it allows the experimenter to learn about some fundamental features about waves, plasmas, and plasma experiments. Much is known already about these waves, but not all. This dearth of knowledge is most evident when we move into the arena of non-linear effects, where just the tips of the icebergs are being explored. At least in these days, we know there are icebergs! Let us summarize some wave–plasma experimental techniques in what follows. Four questions will be treated: how will a wave be excited, how will it propagate, and how can it be detected? Finally, what conclusions can be obtained?

A typical experiment setup for ion-acoustic wave studies is shown in Figure 13.12. The wave is launched by the local creation of an ion density perturbation. Such a density perturbation can be created by a local additional ionization caused by a laser or from a grid structure or a metallic plate to which a voltage signal is applied. The latter techniques are easy, and the basic excitation mechanism is now becoming understood.

For example, imagine that the potential of a grid that is to be used as a launching structure were initially negative. Ions from the right or the left will freely gain energy as they approach the grid, will pass through the mesh and lose this additional energy as they climb the energy hill on the other side, and will leave with the same energy with which they started. However, if at the time that the ions are at the bottom of the well with the additional increment of energy gained in falling there, the grid voltage is suddenly removed, the ion perturbation will leave the grid region with an additional increment of energy. In fact, its velocity can be computed by equating the kinetic energy with the potential energy that was gained by the ions as they fell into the well:

$$\frac{1}{2}M_i v_i^2 = q \, \Delta\phi$$

or

$$v_i = \sqrt{\frac{2q \, \Delta\phi}{M_i}} \tag{81}$$

This potential well created by the grid is shown in Figure 13.13. This free-streaming burst of ions is not a wave! However, the experimenter usually just detects perturbations in, say, an ion density, and this anomalous signal may be interpreted as a wave. No collective effects are involved, and this perturbation is sometimes called a pseudowave. Now let us assume that the velocity of this ion perturbation is approximately $2 \times c_s$. It can be shown that it will then launch an ion-acoustic wave through what is called a beam–plasma instability.

Another technique to launch an ion density perturbation that will evolve into an ion-acoustic wave is to use a double-plasma machine. The double-plasma

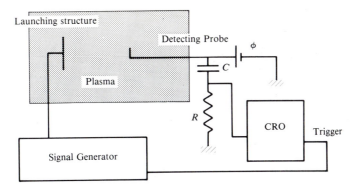

Figure 13.12   Experimental setup for plasma wave experiment. The detecting probe is movable.

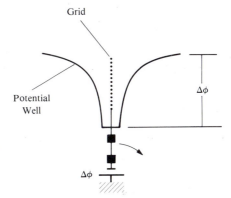

Figure 13.13   Negative potential $\Delta\phi$ that is turned on at $t = -\infty$ creates a potential well about the grid. This potential is removed at $t = 0$, causing ions to leave the grid region.

machine is a novel device that has been found to be very useful. Two independent plasmas are created in juxtaposed chambers. A very negatively biased screen separates the two chambers as shown in Figure 13.14. The negatively biased screen prevents electrons from communicating from one chamber to the other. If the plasma potentials of the two chambers are adjusted to be equal, then ions will also not traverse from one chamber to the other. If we suddenly increase the potential of one of the chambers (the *driver*), a burst of ions will go to the other chamber (the *target*). Again, an ion wave can be excited.

Now the wave is excited. We had the choice of exciting a cw signal, a sine-wave burst, or a pulse. We will choose one of the latter two since then it is easy to separate any electrostatic coupling (electromagnetic wave) from the propagating wave. The velocities of the two are very different (3×10^{10} cm/s verses $\approx 3 \times 10^5$ cm/s for the ion-acoustic wave), so one will appear instantaneously on the scope while the other will be delayed some tens of microseconds. How do we detect the wave?

The simplest method is to use the same Langmuir probe that was used to diagnose the plasma originally. If we bias it above (beneath) the plasma potential, then we will detect perturbations in the electron (ion) saturation current. The first

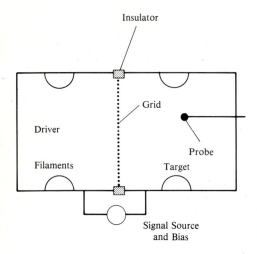

Figure 13.14 Double-plasma machine. The filaments are biased at approximately − 50 V with respect to the chamber, which serves as the anode. Frequently, a multidipole magnetic structure surrounds the chamber.

case yields data that are $(M_i/m_e)^{1/2}$ larger, so we usually make use of this mass-dependent amplification. Remember with quasi-neutrality that

$$\frac{\Delta n_e}{n_e} \approx \frac{\Delta n_i}{n_i} \approx \frac{q\,\Delta\phi}{k_B T_e} \tag{82}$$

Typical sketches of data are shown in Figure 13.15. The data were taken at two locations in space. Note that the electrostatic coupling suffers no time delay in the time scale of 100 μs, which is depicted in this figure, since it propagates at the velocity of light. The two positions of the probe were separated by a few centimeters.

A summary of data taken at various locations and times is given in Figure 13.16. In addition, the results using two different gases are also shown. The velocity can be calculated by taking the slopes of the indicated straight lines. We

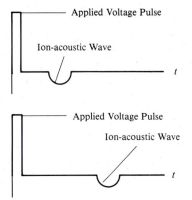

Figure 13.15 Detected signals on the probe at two distances from the launching structure.

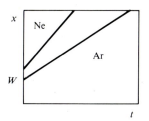

Figure 13.16 Trajectories of the ion-acoustic wave in plasmas containing different ion species. The $t = 0$ intercept yields the transient sheath W obtained for plate excitation.

find that the velocity is given by

$$c_s = \sqrt{\frac{k_B T_e}{M_i}} \tag{83}$$

There may be some damping due to collisions, plasma inhomogeneities, or wave–particle interaction (Landau damping), but these can be considered to be beyond the scope of this text.

What have we done? We have looked at linear ion-acoustic waves. You are probably wondering what would happen if we started all this work over again with the nonlinear fluid equations germane to the ion-acoustic wave propagation, let $\Delta n_{e\,or\,i}$ be on the order of $n_{e\,or\,i}$, and carried out the derivation of the wave equation. We would obtain the Korteweg–de Vries (KdV) equation

$$\frac{\partial \Delta \phi}{dt} + a\, \Delta \phi \frac{\partial \Delta \phi}{\partial x} + b \frac{\partial^3 \Delta \phi}{\partial x^3} = 0 \tag{84}$$

This equation was introduced in Chapter 11. Considerable attention is currently being paid to predicting and experimentally verifying the soliton solutions to this equation and higher-dimensional generalizations of it.

XI. Conclusion

The reader has been introduced to some of the equations of plasma physics, with a particular emphasis on plasmas that can be easily created in the laboratory. Certain sheaths and diagnostic techniques were outlined. Particular attention was given to a low-frequency wave that could propagate in a plasma. Another wave peculiar to a plasma propagates at frequencies above the electron plasma frequency. At these frequencies, the ions can be considered to be stationary, and the derivation of some of their characteristics is left for the problems. If we study the plasma in a magnetic field environment, the variables are all vectors and the analysis is considerably more difficult. There are several rewards at the end as several new plasma effects are uncovered. This is particularly true if we model the plasma with kinetic theory, refrain from making any linearization assumptions, or follow the motion of several thousands of individual charged particles on a large computer.

A long-term goal among some plasma physicists is to realize Einstein's statement that energy and mass must be conserved. Extracting heavy hydrogen from the oceans is fairly inexpensive. Fusing these deuterium atoms together produces helium and a neutron that weighs less than the original deuterium atoms. This occurs in the sun and in hydrogen bombs. Hence the unlimited energy source is at our doorstep if we can just find the proper key to open the door. Recent advances in superconductivity may facilitate the construction of large and powerful magnets that may assist in making this dream a reality.

PROBLEMS

1. Derive the equation of continuity for electrons by assuming the electrons exist in a box whose dimension is $(\Delta x)A$ if there is a net current out of the box indicating a decrease of density within the box. Show that Equation (1) results in the limit of $\Delta x \to 0$.

2. Find required the modification to Equation (1) if additional particles are created through ionization, $n_{\text{ionization}} \approx \alpha n_e$, and if particles are destroyed by recombination, $n_{\text{recombination}} \approx \beta n_e^2$. Solve this equation in the spatially homogeneous limit $\partial/\partial x = 0$.

3. Calculate the number of particles in a Debye sphere in a plasma where $n_e = 10^{12} \text{ cm}^{-3}$ and $T_e = 100$ eV.

4. Several plasmas in space have widely different densities. Calculate the electron plasma frequency f_{pe} for each.

$$\text{Earth's wake } n_e \approx 10^{-2} \text{ cm}^{-3}$$

$$\text{Ionosphere } n_e \approx 10^6 \text{ cm}^{-3}$$

$$\text{Interior of the sun } n_e \approx 10^{20} \text{ cm}^{-3}$$

5. Assume that there are ν collisions per second between the electron slab and neutral particles as it oscillates about the equilibrium position shown in Figure 13.4. Derive the equation of motion for the electron slab and show that the oscillation will finally decay to zero.

6. Show that the term $[k_B T_e/M_i]^{1/2}$ has the dimensions of a velocity that are L/T, where L is the length and T is the time.

7. Show that (39) follows from (38).

8. Postulate the effect of having a distribution function with a positive slope of $\partial f/\partial v$ at the velocity of a wave. This is sometimes called a "bump-on-the-tail."

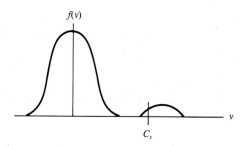

9. Calculate the maximum current density that can be thermally emitted from a tungsten filament whose work function is 4.5 eV. The melting point of tungsten is 3683°K.

10. Nickel coated with various oxides operating at 1100°K has a work function of 1 eV. Calculate the thermally emitted current density.

11. If the space-charge region is 3 mm thick in front of a cathode, calculate the space-charge-limited current if $\phi_0 = 100$ V.

12. Calculate the Larmor radius for ionizing electrons in the multidipole magnetic cage shown in Figure 13.9. These electrons must have sufficient energy to ionize three argon ions whose ionization potential is approximately 15 V. The magnetic field strength is $B = 100$ gauss.

13. Calculate the dimension of a Langmuir sheath in front of a biased metal surface in an argon plasma if -20 V is applied as a bias. The temperature and density of the plasma are $T_e \approx 1$ eV and $n_e \approx 10^8$ cm^{-3}.

14. Calculate the transient sheath dimension if -20 V is suddenly applied to an unbiased metal surface in the plasma described in Problem 13.

15. Show that a Langmuir probe is a more sensitive detector of perturbations in the electron saturation current in a plasma with $T_e \gg T_i$. This is important for wave-detection experiments.

16. A measured Langmuir probe characteristic is shown below. Determine the plasma parameters if the area of the probe is $A \approx 10$ mm^2.

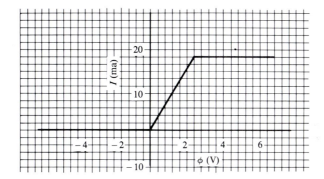

17. If the electron temperature is the same in a plasma, $T_e \approx 1$ eV, plot the expected ion-acoustic velocities for all the noble gases.

18. Plot the trajectories in space and time for an ion-acoustic wave in a plasma if it reflects from a surface located at $x = L$. Work is still being performed to completely understand the reflection mechanism.

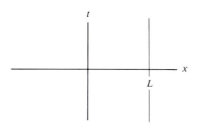

Chapter Fourteen

High-Frequency and
High-Power Tubes

The scientists have taken their long sojourns into the depths of the structure of the atom and on their way out have prepared for us the table of haute cuisine of solid-state devices that have been outlined in the previous chapters. The diner at this table is in a state of awe as the morsels are developed and brought forth with a rapidity that boggles the mind. The conversation quickly focuses on the search for new valleys in many parts of the world in the hope of finding one more fertile than the original silicon valley in order to maintain the riches that are on this table. But as the diner approaches the end of this feast, some troubling questions may be gnawing at the soul. The first revolves around the point, "Can solid-state devices do everything?" and the second raises an equally troubling idea: "Is there any future for the new ideas that also arose during the time of or shortly after the Great Depression that do not include $\hbar$ or those other symbols?" The answers to these two questions can be given as an emphatic "No!" to the first query and "Yes!" to the second. There are times when it is preferable to visit the cafe with the more substantial fare, in particular when the palate desires the high power and high frequency that cannot yet be found in the solid-state establishments.

The concept of velocity modulation will be described initially and tubes that have grown out of this idea will be presented. Also, a tube that is based on a wave–particle interaction will be discussed. As time passes, solid-state devices have replaced many of the low-power applications of these devices (even at frequencies up to 10^{11} Hz with the ballistic transistor today), but heat-dissipation problems impose severe restrictions as the level of the desired output power is increased. Also, conventional vacuum-tube technology does maintain some interesting features, for example, in applications where radiation resistance is important.

I. Velocity Modulation

One area of high-frequency power generation where solid state has not yet made an inroad is the generation of high-power microwaves. At frequencies on the order of 10^{10} Hz and powers in the kilowatt to megawatt range, we still have to resort to the "tried and true" of vacuum tubes. These tubes have peculiar characteristics in that the basic physical mechanism for their operation is **velocity modulation**. Several tubes such as klystrons and magnetrons depend on this mechanism for operation and it will be discussed later. Several of the techniques and ideas that were discussed already in plasmas have their analog in this area and may actually have had their origin here.

In Figure 14.1, we illustrate the basic configuration that allows velocity modulation. Electrons are first "boiled off the cathode" (Richardson–Dushman equation) and are accelerated by the potential source ϕ_0. The average velocity of these electrons is computed from the relation

$$\frac{1}{2}m_e v^2 = q\phi_0 \tag{1}$$

This is similar to the burst of ions that was described in the previous chapter. At a certain location, which we will take as $x = 0$, a pair of closely separated grids is located. A sinusoidal voltage generator is connected between the grids, and the electrons will gain an increment of energy $q\,\Delta\phi = q\phi_1 \sin \omega t$. Therefore, the total energy that these electrons will have in the drift space $x > 0$ is given by

$$\frac{1}{2}m_e v^2 = q(\phi_0 + \phi_1 \sin \omega t) \tag{2}$$

We can solve for the velocity of the electrons v as follows:

$$v = \sqrt{\frac{2q\phi_0}{m_e}} \sqrt{1 + \frac{\phi_1}{\phi_0} \sin \omega t}$$

$$\approx v_0 \left[1 + \frac{\phi_1}{2\phi_0} \sin \omega t \right] \tag{3}$$

Note that at certain times this velocity will be greater than its average value v_0, and at some times it will be less (i.e., the velocity is modulated with the voltage ϕ_1).

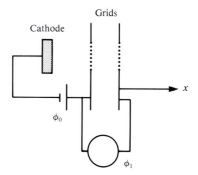

Figure 14.1   Cathode and grid pair that causes velocity modulation.

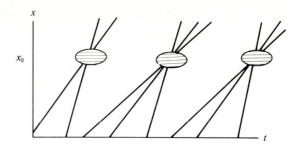

Figure 14.2 Applegate diagram describing the electron trajectories in a velocity-modulated tube.

In Figure 14.2, the trajectories of the electrons in space and time, which are now freely streaming away from the modulating pair of grids, are given. The slope of each line is given by $dx/dt = v$. This diagram is sometimes called an Applegate diagram. Note that at certain locations in space the trajectories have "bunched together," which indicates that there will be a larger local electron density there. This is more clearly shown in Figure 14.3, where the density of the electrons increases up to the point of bunching. The total number of electrons N is assumed to be a constant because there is no ionization or loss mechanism in the vacuum drift space. Hence the current density, given by

$$j_{\text{local}} \approx n_{\text{local}} q v_0$$

will also be larger! If the density of electrons passing through the grids is n_0, then the signal perturbation will be enhanced by an amount

$$\frac{j_{\text{local}}}{j_0} \approx \frac{n_{\text{local}} q v_0}{n_0 q v_0} = \frac{n_{\text{local}}}{n_0} \tag{4}$$

This means that if another pair of grids were placed at this location, $x = x_0$, the high-frequency signal will be *amplified*. We can also think of this oscillating

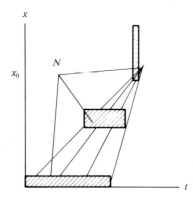

Figure 14.3 Schematic representation of the bunching of electrons in space and time. The total number N of electrons is a constant.

current as a source term for Maxwell's equations if we want a more formal approach.

Example 14.1   Estimate the separation distance ΔL between the grids located at x_0 if the frequency of operation is f.

Answer: The passage of the "modulated" electron bunch at the location x_0 requires a time $T \approx \Delta L / v_0$ to pass through the grid region. This acts as one half-cycle of the oscillating current and corresponds to a time $\approx \frac{1}{2} f$. Therefore, $\frac{1}{2} f \approx \Delta L / v_0$ or $\Delta L \approx v_0 / 2f$.

II. Klystrons

In the discussion of velocity modulation given previously, it was suggested that amplification caused by the velocity-modulated electrons could be employed to excite a second cavity, as shown in Figure 14.4. The application of a two-cavity klystron as an amplifier is straightforward. From Figures 14.2 and 14.3, we note that the electron perturbation bunches at a fixed location and that there is an amplification as given in (4). The detailed electromagnetic field structure in the cavities is complicated, although it is possible to present a simplified analysis of a frequently used cavity.

It is common to employ *reentrant* cavities as shown in Figure 14.5. The modulated electron beam passes through the hole in the center of the cavity. As

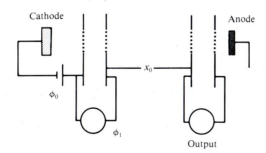

Figure 14.4   Two-cavity klystron.

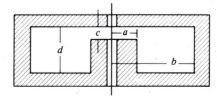

Figure 14.5   Cross-sectional view of a reentrant cavity.

such, the current density **j** will be in the same direction as the electric field **E**, which will be confined to the region $r < a$.

The resonant frequency ω_r of this cavity can be approximately given by

$$\omega_r \approx \frac{1}{\sqrt{LC}} \tag{5}$$

where

$$C \approx \frac{\varepsilon_0 \pi a^2}{c} \quad \text{and} \quad L \approx \mu_0 \frac{d}{2\pi} \ln \frac{b}{a}$$

which are approximated by assuming that the narrow gap region is a parallel plate capacitor and the coaxial region is similar to a coaxial cable. Correction terms should be included to account for stray capacitances, but this simple model is a reasonable place to start.

The two-cavity klystron can also be used as an oscillator by externally coupling a portion of the signal from the output cavity back into the initial bunching cavity. If this "coupled signal" has the proper phase relationship, it will provide the necessary negative feedback to cause the tube to oscillate. This feedback can also be provided internally by returning the modulated electron beam back into the cavity with a repeller voltage ϕ_r connected to a plate as shown in Figure 14.6. If the distance between the modulating cavity and the repelling electrode is L, this feedback can occur for several different values of repeller voltage and thus several modes of operation are possible. This klystron is called a *reflex* klystron.

There are two methods of tuning the frequency of a reflex klystron. The first is to mechanically change the separation of the two parallel plates of the capacitive element of the reentrant cavity. The second is to adjust the repeller voltage ϕ_r, which will control the phase and net transit time of the velocity bunching electrons. This latter tuning will also affect the output power.

Both types of klystrons have found wide application in high-power radar transmitters. It is not uncommon in high-power applications to use several stages of bunching cavities before the desired output power levels are achieved. Hence the physical size of some tubes can be the size of a human.

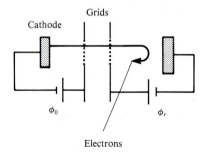

Figure 14.6 Reflex klystron. The path traveled by the bunching electrons is such that they actually bunch in the cavity.

III. Magnetrons

In both of the tubes discussed previously, the electrons drift in straight lines. If we cause the electrons to drift across a magnetic field, the orbits will be curved. In fact, the orbits can become closed circles. In Figure 14.7, a typical magnetron is shown. It consists of an electron-emitting cathode at the center surrounded by a positively biased anode structure. As the anode is on the outside, it is common to ground it and bias the cathode negatively for safety considerations. The trajectories of the electrons that are emitted from the central cathode are accelerated to the outer anode by the voltage ϕ_0. This electron stream, which may have a small thermal spread in its velocity about an average value, is bent due to the magnetic field. With a proper choice of magnetic field and accelerating voltage, these electrons will orbit about the cathode as almost a steady circular stream.

Surrounding the drift region in Figure 14.7 are four small cavities with a small gap region facing the drifting electron stream. Due to the random fluctuations of the streaming electrons, a small oscillating electric field will be created in the narrow gaps. This electric field will cause the necessary "velocity modulation" of the electrons as they stream about the central cathode. The high-frequency electromagnetic energy is coupled from the magnetron through one of these small cavities.

The precise value of the radial coordinate where these electrons reside can be computed from the following argument. Let the radius of the cathode be a and the radius of the anode be b. The potential at any radius r that is in the drift region is given by

$$\phi(r) = \phi_0 \left[\frac{\ln (r/a)}{\ln (b/a)} \right] \tag{6}$$

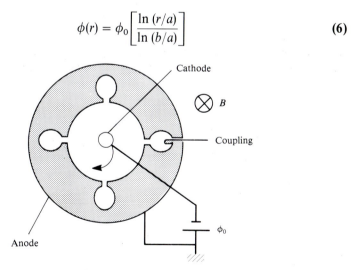

Figure 14.7 Electrons emitted from the center cathode are accelerated to the outer anode. Their trajectories are curved due to the magnetic field in a magnetron.

and the radial velocity of an electron at this point is given by

$$v(r) = \sqrt{\frac{2q\phi(r)}{m_e}} \qquad (7)$$

At this radius r, the electron will experience three forces:

1. Outward centrifugal force, mv^2/r
2. Outward electrical force, qE_r
3. Inward magnetic force, qvB

In equilibrium, we write

$$\frac{mv^2}{r} + \frac{q\phi_0}{r[\ln(b/a)]} = qvB \qquad (8)$$

Hence, the radius r can be determined.

The tubes that have been mentioned here can have powers on the order of kilowatts with wavelengths in the millimeter to tens of centimeter range. A 2.4-GHz (10^9-Hz) magnetron with a power of approximately 500 watts may have cooked your dinner while several 200-kW, 70-GHz microwave sources may heat a plasma fusion machine in the future. At the present time, solid-state devices are not useful in these ranges due in part to heat-dissipation problems.

IV. Gyrotrons

Recent demands by the controlled thermonuclear fusion community to develop a high-power source ($\approx$ hundreds of kilowatts) at extremely short wavelengths (a few millimeters) in order to heat a plasma up to temperatures where the heavy hydrogen ions would fuse together have led to the development of a new class of microwave tube called the gyrotron. To understand the operation of the gyrotron, consider the model shown in Figure 14.8.

Electrons are emitted from the cathode and are accelerated with the voltage ϕ_0; they enter the ring electrode region with a longitudinal velocity

$$v_b = \sqrt{\frac{2q\phi_0}{m_e}} \qquad (9)$$

At the ring electrode that surrounds the incoming electron beam, these electrons encounter a transverse electric field $\mathbf{E_T}$ created by the potential ϕ_D. The magnitude of $\mathbf{E_T}$ is much larger than that caused by the negative charge of the electron beam. This causes the electrons to be accelerated in the transverse direction. An external magnetic field that increases in strength in the direction of the electron motion causes these electrons to gyrate about the longitudinal axis.

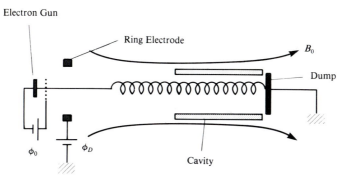

Figure 14.8   Schematic representation of a gyrotron.

As noted in Chapter 12, the Larmor radius of these gyrating electrons is given by

$$r_L = \frac{m_e v_e}{qB} \qquad (10)$$

where v_e is taken to be the transverse velocity of the electrons.

The electrons, as they pass from the weaker magnetic field into the region of stronger magnetic field, will have a decreasing Larmor radius. A detailed mathematical description is difficult, but it is possible to draw an analogy to a figure skater who starts a finale with arms outstretched and gradually increases in rotational velocity as the arms are brought close to the body. Rotating electrons also increase in velocity such that relativistic corrections may have to be incorporated into the correct description of a gyrotron.

The gyrating electrons pass into a cylindrical microwave cavity whose electric field of the TE (transverse electric field components only) mode interacts and velocity modulates these electrons. If this is done, we find that there is an interaction between the rotating electrons and the cavity electric field, a transfer of energy from the electrons to the field, and hence an amplification of the field. Two possible mode structures for the cavity are shown in Figure 14.9.

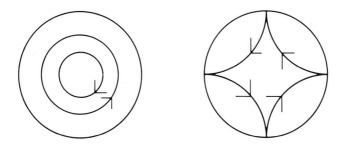

Figure 14.9   Two possible modes for the gyrotron cavity. Arrows indicate the direction of the electric field, which is entirely in the transverse plane.

At first glance, the gyrotron appears to be somewhat similar to a magnetron in that the velocity modulation of the electrons occurs in the circumferential drift. A major difference is that there are no externally machined cavities, as are shown in Figure 14.7.

Example 14.2   Estimate the dimensions of the cavity if we want to design a gyrotron to operate at 60 GHz.

Answer: The radius where the electric field is the maximum is approximately $r_0/2$, where r_0 is the radius of the cavity. The velocity of the rotating electrons is approximately equal to the velocity of light c. Therefore, if we assume that 10 free-space wavelengths are required to bunch the electrons and set this distance equal to the circumference at this radius, we write

$$2\pi \frac{r_0}{2} \approx 10\lambda \approx 10\frac{c}{f} \quad \text{or} \quad r_0 \approx \frac{10}{\pi}\frac{c}{f} \approx 1.6 \text{ cm}$$

It has been found that the efficiency of a gyrotron can be on the order of 50%. As such, its applicability for high-power, high-frequency operation is presently being explored by several communities in several nations.

V. Further Comments on Velocity Modulation

The three high-frequency tubes that have been qualitatively described, the klystron, the magnetron, and the gyrotron, have all relied on the physical effect of velocity modulation. The subsequent bunching of the electrons in space was interpreted using an Applegate diagram. In this section, velocity modulation will be reexamined in terms of a distribution function in velocity and configuration space (i.e., in terms of the phase space that was described in Chapter 4).

The equation that governs the evolution of the distribution function in phase space is the kinetic equation

$$\frac{\partial f}{\partial t} + \frac{\partial (fv)}{\partial x} + \frac{\partial [f(F/m)]}{\partial v} = \left(\frac{df}{dt}\right)_{\text{collisions}} \tag{11}$$

where the term F/m is taken to be the self-consistent electric field between the electrons and v has replaced $\langle v \rangle$ for notational purposes. The density of the electrons is computed in terms of the distribution function by the integral

$$n = \int_{-\infty}^{\infty} f \, dv \tag{12}$$

In Chapter 4, the evolution of the distribution function into the Maxwell–Boltzmann distribution, which is the state of thermodynamic equilibrium, was considered. Here we will examine how a particular distribution that is germane to the concept of velocity modulation will evolve. This distribution is the "water-bag" distribution.

The philosophy of the water-bag model is to assume that the distribution f is localized in some region in phase space initially and is zero elsewhere and, furthermore, that the self-consistent electric field is *small* and the effects of it can therefore be neglected. With this approximation, the distribution function evolves according to

$$\frac{\partial f}{\partial t} + v\frac{\partial f}{\partial x} = 0 \tag{13}$$

This states that the distribution function will just translate in space and time with the particular velocity v that it had at $x = 0$.

It is possible to approximate the initial distribution function for the *velocity-modulated* electrons (the velocity is modulated with a sinusoidal modulation that has a peak amplitude v_0) that are between a pair of modulating grids that are assumed to be at $x = 0$ as

$$f(x = 0, v, t) = \frac{n_0}{2\,\Delta v} P\left[\frac{v - v_b - v_0\sin\omega t}{\Delta v}\right] \tag{14}$$

In writing (14), we have assumed that the electrons emitted from a hot cathode have a spread in velocity equal to $\pm\Delta v$ about an average value and the beam velocity is given by v_b. The maximum velocity of the ac modulation is v_0. The function $P(\mu)$ is the pulse function, which is defined as

$$P(\mu) = \begin{cases} 1, & |\mu| \le 1 \\ 0, & |\mu| > 1 \end{cases} \tag{15}$$

The electron density at any location x can be computed from the integral given in (12), which is rewritten as

$$n = \int_{-\infty}^{\infty} f(x, v, t)\, dv \tag{16}$$

At $x = 0$, we compute

$$n(x = 0) = \int_{-\Delta v}^{\Delta v} \left[\frac{n_0}{2\,\Delta v}\right] dv = n_0 \tag{17}$$

which states that the density is constant and equal to n_0.

The distribution function will evolve according to the simplified Vlasov equation (13), and the solution can be written as

$$f(x, v, t) = \frac{n_0}{2\,\Delta v} P\left[\frac{v - v_b - v_0 \sin\{(\omega/v)(x - vt)\}}{\Delta v}\right] \tag{18}$$

The evolution of this distribution function in space is shown in Figure 14.10. The spatial axis is given in terms of the beam velocity v_b as $x = v_b t$. Certain general features can be gleaned from this figure. First, the perturbation is translated in space at an average velocity given by the beam velocity v_b. Second, each individual particle will possess a distinct local velocity component that could add or subtract from this average velocity. The temporally later launched particles in a particular cycle have a greater velocity than this average, and those launched earlier have a velocity that is less than this average. The electrons will eventually all meet at the same location in space at the same time, a process called *bunching*. From the phase-space diagram in Figure 14.10, we note that the peaks of the local velocity are either accelerated or retarded by one quarter-cycle in the space between $x = 0$ and $x = x_{\text{bunch}}$, which is the location where they first bunch.

The boundaries separating the region where $f = n_0/2\,\Delta v$ and $f = 0$ are determined by

$$v - v_b + v_0 \sin\left\{\left(\frac{\omega}{v}\right)(x - vt)\right\} = \pm\Delta v \tag{19}$$

Since the distribution function is a constant between these boundaries and zero elsewhere, the characteristics of the distribution function can be determined just from an examination of the evolution of these boundaries. It is possible to predict

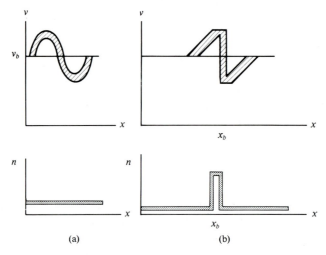

(a)   (b)

Figure 14.10 Velocity distributions and electron density at (a) $t = 0$ and (b) $t = t_b$.

the location of the bunching by determining the location where the phase-space diagram has steepened. This location of bunching will occur at the location where

$$\frac{\partial v}{\partial x} \to -\infty$$

This can be found from (19) as

$$\frac{\partial v}{\partial x} = -\frac{(\omega/v)v_0 \cos\{(\omega/v)(x - vt)\}}{1 - (\omega x/v^2)v_0 \cos\{(\omega/v)(x - vt)\}} \tag{20}$$

The steepening will occur where the denominator of (20) is equal to zero.

The multiplicity of locations in the Applegate diagram where this bunching occurs can be emphasized if we set

$$\frac{\omega x v_0}{v^2} = 1 \tag{21}$$

These locations in x and in time t are then given by the solutions of

$$\cos\left\{\left(\frac{\omega}{v}\right)(x - vt)\right\} = 1 \tag{22}$$

The first time $t = t_0$ that this occurs is when

$$x - vt = 0$$

or

$$t_0 \approx \frac{x_b}{v_b} \approx \frac{v_b}{\omega v_0} \tag{23}$$

where (21) has been included at the location of bunching at $x = x_b$.

The total charge ΔN_{bunch} that has accumulated about the point x_b can be computed as follows. From (21), we write

$$x_b \pm \Delta x = \frac{v^2}{\omega v_0} = \frac{(v_b \pm \Delta v)^2}{\omega v_0} \approx \frac{v_b}{\omega v_0}(v_b \pm 2\,\Delta v) \tag{24}$$

The total charge is found from the integral of the distribution function over all velocities and over the region where the electrons have bunched. We write

$$\Delta N_{\text{bunch}} = \int_{x_b - \Delta x}^{x_b + \Delta x} dx \int_{-v_0}^{v_0} f\, dv = \frac{4 n_0 v_b}{\omega} \tag{25}$$

where the distribution function in the undisturbed region

$$f = \frac{n_0}{2\,\Delta v}$$

has been employed. Similarly, the density in the beam in the same region of space if there is no velocity modulation is given by

$$N_{\text{unbunched}} = \left(\frac{4\,\Delta v v_b}{\omega v_0} \right) n_0 \tag{26}$$

The densities corresponding to these times are also shown in Figure 14.10. These particles are moving with approximately the beam velocity v_b and therefore act as a source current for a second pair of grids or a cavity located at the location $x = x_{\text{bunch}}$. The ratio of (25) and (26) is a measure of the amplification caused by the velocity modulation, and this ratio is $v_0/\Delta v$. This is similar to the amplification ratio postulated in (4).

This formal treatment of velocity modulation has received considerable attention in plasma physics calculations recently. The origin of the name "water bag" to describe the distribution function that has been employed here and in these calculations can be thought of as follows: As noted in Figure 14.10, the phase-space distribution boundaries were distorted as time progressed. However, the total density that was the integral over all space remained the same. This would be similar to a fixed amount of water filling a plastic bag, which would then be distorted. Unless the bag burst, no water would appear outside the boundaries defined by the plastic.

A simple physical model for velocity modulation can be found at the seashore. Waves that are far away in the sea have a small amplitude. As they approach the beach, their amplitude gradually increases and their width decreases until nonlinear overtaking occurs.

VI. Space-Charge Waves

In addition to the concept of velocity modulation discussed previously, an additional effect will be useful for the amplification and generation of high-frequency waves. It is based on the fact that a beam of charged particles can support waves. Recall that in Chapter 13 we examined the details of an ion-acoustic wave propagating in a plasma that was stationary and had no steady-state velocity. At this time, we will remove that stationary restriction and find that a *space-charge* wave can be found for this system. Based on practical considerations, it is more prudent to examine an electron beam rather than an ion beam, as microwave tubes that are based on this idea operate at frequencies that are commensurate with the electron motion. It will also be advantageous to use the ∇ operator, as two-dimensional effects will be later included.

The starting point in the analysis is the equation of continuity

$$\frac{\partial \rho}{\partial t} + \nabla \cdot \mathbf{j} = 0 \tag{27}$$

where the magnitude of the current density $\mathbf{j}$ is given by

$$j + j_0 = (\rho + \rho_0)(v + v_0) \tag{28}$$

where the subscript 0 indicates a dc term and the terms without the subscript are the fluctuating terms. Also, it will be assumed that the fluctuating terms are small, so (28) can be written as

$$j + j_0 \approx (\rho v_0 + \rho_0 v) + \rho_0 v_0 \tag{29}$$

where the term containing the product of two fluctuating terms is neglected.
Maxwell's equations in free space are written as

$$\nabla \times \mathbf{E} = -\mu_0 \frac{\partial \mathbf{H}}{\partial t} \tag{30}$$

$$\nabla \times \mathbf{H} = \mathbf{j} + \varepsilon_0 \frac{\partial \mathbf{E}}{\partial t} \tag{31}$$

$$\nabla \cdot \mathbf{E} = \frac{\rho}{\varepsilon_0} \tag{32}$$

$$\nabla \cdot \mathbf{H} = 0 \tag{33}$$

From these equations, we can derive a wave equation as follows:

$$\nabla \times \nabla \times \mathbf{E} = -\mu_0 \frac{\partial (\nabla \times \mathbf{H})}{\partial t} = -\mu_0 \frac{\partial [\mathbf{j} + \varepsilon_0 (\partial \mathbf{E}/\partial t)]}{\partial t}$$

$$\nabla^2 \mathbf{E} - \nabla(\nabla \cdot \mathbf{E}) - \frac{1}{c^2} \frac{\partial^2 \mathbf{E}}{\partial t^2} - \mu_0 \frac{\partial \mathbf{j}}{\partial t} = 0$$

or finally

$$\nabla^2 \mathbf{E} - \frac{1}{c^2} \frac{\partial^2 \mathbf{E}}{\partial t^2} = \frac{1}{\varepsilon_0} \nabla \rho + \mu_0 \frac{\partial \mathbf{j}}{\partial t} \tag{34}$$

In (34),

$$c^2 = \frac{1}{\mu_0 \varepsilon_0}$$

where c is the velocity of light in a vacuum. Note that (34) is a standard wave equation with two source terms on the right side. It specifies the electric field resulting from the distribution of charges and currents.

The problem that remains in order to solve (34) is that we must be able to specify the current and the charge density in terms of the electric field $\mathbf{E}$. This can be done using the equation of motion

$$\frac{\partial \mathbf{v}}{\partial t} + \mathbf{v} \cdot \nabla \mathbf{v} = -\frac{q}{m_e} \mathbf{E} \tag{35}$$

In writing (35), we have neglected any contribution to the force due to any magnetic fields. The magnetic field will later be assumed to be sufficiently strong such that only one-dimensional motion will be allowed, and the self-induced magnetic field will be neglected since v/c will be considered to be small.

Let us assume that an electron beam has a dc velocity that is directed in the $\mathbf{u}_z$ direction. Again neglecting terms that contain the product of fluctuating terms, (35) can be written as

$$\frac{\partial \mathbf{v}}{\partial t} + v_0 \frac{\partial \mathbf{v}}{\partial z} = -\frac{q}{m_e} \mathbf{E} \tag{36}$$

We next make the approximation that a wave exists on this beam and has a temporal and spatial dependence of the form $e^{-i(\omega t - \beta z)}$. This will simplify the derivatives since

$$\frac{\partial}{\partial t} \to -i\omega \quad \text{and} \quad \frac{\partial}{\partial z} \to i\beta$$

Employing these definitions, (36) becomes

$$\frac{q}{m_e} \mathbf{E} = i(\omega - \beta v_0)\mathbf{v} \tag{37}$$

and (34) takes the form

$$\nabla^2 \mathbf{E} + k^2 \mathbf{E} = \frac{1}{\varepsilon_0} \nabla \rho - i\omega \mu_0 \mathbf{j} \tag{38}$$

where the free-space wave number $k = \omega/c = 2\pi/\lambda$ and λ is the free space wavelength.

It is now possible to eliminate ρ and $\mathbf{j}$ from (38) by assuming that the fluctuating motion of the charge is limited to the z direction. This is true for the two limiting cases of no magnetic field in an infinite system, where the transverse perturbations will average to zero, and in a strong magnetic field oriented in the direction of the electron motion, where there is no transverse motion. In these two limits, (37) becomes

$$\frac{q}{m_e} E_z = i(\omega - \beta v_0)v_z \tag{39}$$

and (27) is written as

$$\beta j_z + \omega \rho = 0 \tag{40}$$

The fluctuating part of the current density given in (39) will also only have a z component, which can be stated as

$$j_z = \rho_0 v_z + \rho v_0 \tag{41}$$

From (40) and (41), we obtain

$$\rho = \frac{-\beta \rho_0}{\omega - \beta v_0} v_z \tag{42}$$

Eliminate v_z between (39) and (42) and write

$$\rho = \frac{-(q/m_e)\beta \rho_0}{i(\omega - \beta v_0)^2} E_z \tag{43}$$

Therefore, (38) becomes

$$\nabla^2 \mathbf{E} + k^2 \mathbf{E} = \frac{1}{\varepsilon_0} \nabla \left[\left\{ \frac{-(q/m_e)\beta \rho_0}{i(\omega - \beta v_0)^2} \right\} E_z \right] - i\omega \mu_0 \frac{\omega}{\beta} \left\{ \frac{-(q/m_e)\beta \rho_0}{i(\omega - \beta v_0)^2} \right\} E_z \mathbf{u}_z$$

where $\mathbf{u}_z$ is a unit vector. Using the electron plasma frequency, which is defined as

$$\omega_p^2 = \frac{\rho_0 [-(q/m_e)]}{\varepsilon_0}$$

this equation can be written as

$$\nabla^2 \mathbf{E} + k^2 \mathbf{E} = \frac{\omega_p^2 \beta}{i(\omega - \beta v_0)^2} \nabla E_z + \left(\frac{\omega}{c}\right)^2 \frac{\omega_p^2}{(\omega - \beta v_0)^2} E_z \mathbf{u}_z \tag{44}$$

The problem can be simplified if only the z component of the electric field is considered. If this is done, then (44) becomes a scalar equation:

$$\nabla^2 E_z + \left\{ k^2 + \omega_p^2 \left[\frac{\beta^2 - (\omega/c)^2}{(\omega - \beta v_0)^2} \right] \right\} E_z = 0 \tag{45}$$

At this stage, we have not yet made any statement about the Laplacian operator ∇^2. Let us use the substitution that it can be broken up into two parts and written in rectangular coordinates as

$$\nabla^2 \to \frac{\partial^2}{\partial x^2} + \frac{\partial^2}{\partial y^2} - \beta^2 \tag{46}$$

Equation (45) then becomes

$$\frac{\partial^2 E_z}{\partial x^2} + \frac{\partial^2 E_z}{\partial y^2} + \Gamma^2 E_z = 0 \tag{47}$$

where

$$\Gamma^2 = k^2 - \beta^2 + \omega_p^2 \left[\frac{\beta^2 - (\omega/c)^2}{(\omega - \beta v_0)^2} \right] \tag{48}$$

It is convenient to simplify (48) by defining

$$\beta_p = \frac{\omega_p}{v_0} \quad \text{and} \quad \beta_e = \frac{\omega}{v_0}$$

whereupon (48) can be written as

$$\Gamma^2 = (\beta^2 - k^2)\left(\frac{\beta_p^2}{(\beta_e - \beta)^2} - 1 \right) \tag{49}$$

If the electron beam were infinite in transverse dimension such that the transverse derivatives are zero, then (37) can only be satisfied if

$$\Gamma^2 = (\beta^2 - k^2)\left(\frac{\beta_p^2}{(\beta_e - \beta)^2} - 1 \right) = 0 \tag{50}$$

There are four roots to this equation if we solve for the propagation constant β. There are two vacuum solutions given by

$$\beta = \pm k = \pm \frac{\omega}{c} \tag{51}$$

and two space-charge solutions given by

$$\beta = \beta_e \pm \beta_p \tag{52}$$

The phase velocity of a space-charge wave is given by

$$v_\phi = \frac{\omega}{\beta} = \frac{\omega}{\beta_e \pm \beta_p} = \frac{\omega}{(\omega/v_0) \pm (\omega_p/v_0)} = \frac{\omega}{\omega \pm \omega_p} v_0 \tag{53}$$

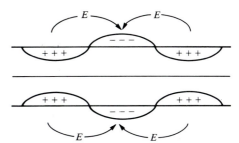

Figure 14.11 Accumulation or depletion of electrons creates local longitudinal electric fields in a space-charge wave.

Hence two space-charge waves propagate with a velocity approximately equal to the beam velocity, one slightly faster and one slightly slower than v_0.

The inclusion of a finite transverse dimension with a large magnetic field does not change these conclusions. The transverse derivatives can be replaced by a constant that is determined by the size of the beam and the boundary conditions. In Figure 14.11, the characteristics of the space-charge waves are shown for a finite-diameter electron beam.

Let us now introduce a second electron beam into the same region as the first beam. Let this second beam have a slightly different velocity from the first. An analysis of the type presented here can be performed on the space-charge waves that propagate on this beam by neglecting the first beam. However, the first beam is present, and now four possible waves are traveling at approximately the same velocity. This will enhance the bunching of the electrons that will occur as the beam propagates from one end to the other, as there will be a continuous coupling between the waves. The enhanced bunching will cause one of the waves to grow in space at the expense of the other.

Another possible scenario is to couple an electromagnetic wave whose velocity in the direction of the electron beam has been slowed down to be approximately equal to the electron beam velocity. The most common technique to slow the electromagnetic wave to this slower velocity is to use a helix structure as shown in Figure 14.12.

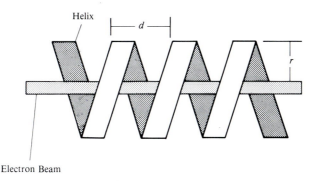

Figure 14.12 Electromagnetic wave is guided by the helix structure and has a velocity in the direction of the beam approximately equal to $(d/2\pi r)c$.

The electromagnetic wave is guided along the helix structure, where it propagates with the velocity of light c. However, since this path length ($\approx 2\pi r$) is longer than the distance d, the net velocity in the direction of the electron beam is reduced to have a value of approximately $(d/2\pi r)c$. A component of the electric field will be in the direction of the electron beam, so it will interact with the beam. The traveling-wave tube (TWT) has very broad band frequency characteristics of amplification. If a portion of the output energy is fed back to the input, the tube will oscillate. This feedback can either be done externally or by exciting a wave whose group and phase velocities are in the opposite directions. This tube is called a backward-wave oscillator (BWO).

VII. Conclusion

Several high-frequency microwave tubes are based on the bunching of electrons that exist in an electron beam that is directed through a velocity-modulating region. This modulation can exist in just a local region of space or be distributed in space. In the latter case, there will be a continuous bunching. If there were neutralizing ions in the region, we would obtain a wave plasma interaction, a general topic of current interest among several plasma physicists.

PROBLEMS

1. Velocity modulation can also be demonstrated by calculating the impulse that an electron receives as it passes through the grids. This can be shown using the integral

$$\int_{mv_\alpha}^{mv} d(mv) = \int_0^t F\, dt$$

   which is the integral of the equation of motion. In this equation, the electrons have an initial velocity v_0 at a time $t = 0$. Find the expression for the electron velocity if a sinusoidal modulating voltage is applied to the bunching cavity.

2. If the cavity in Problem 1 were at $x = 0$, find the particle's position as a function of time.

3. Velocity-modulated charged particles can also be found in wave experiments using a single grid. Show that the appropriate equation of motion that would model ion motion around a Debye shielded grid can be given by

$$\frac{d^2x}{dt^2} = \frac{-x}{|x|} \Phi_0 e^{-|x|} \sin(\omega t + \theta)$$

   where normalized variables have been used. Find the normalizations.

4. It is possible to numerically solve the equation of motion presented in Problem 3; do it and present the results in terms of an Applegate diagram.

5. Show that higher-frequency harmonics of the fundamental of the bunching electron beam depicted in Figure 14.3 have a higher amplitude as the bunching occurs.

6. The reentrant cavity depicted in Figure 14.5 can be used for diagnostics of materials with a relative dielectric constant $\varepsilon_r \neq 1$. This can be accomplished by inserting the dielectric in the cavity in the region $r < a$ and measuring the change of the resonant frequency $\Delta\omega$, and relating it to $\varepsilon_r = \varepsilon_0 + \Delta\varepsilon$. Show this if the dielectric fills the region $r \leq a$.

7. Repeat Problem 6 if the dielectric is a cylindrical rod whose radius $R < a$.

8. Multicavity velocity modulation occurs in klystrons and magnetrons. Using an Applegate diagram, explain the operation of such a multicavity tube. Should there be any particular phase relation between the modulating voltages in each cavity?

9. Let space-charge waves be confined between two infinite parallel metal plates at $x = \pm x_0$. Solve Equation (47) for this case and sketch the dispersion relation β versus ω.

10. Find the phase and group velocities for space-charge waves propagating in an infinite medium $(\partial/\partial x = \partial/\partial y = 0)$.

Appendix A

Self-Similar Solution of Partial Differential Equations

In the study of the charge migration either through transport or diffusion, we frequently encountered linear or nonlinear partial differential equations that had to be solved. Rather than immediately rushing to a computer to obtain a solution, we will describe a powerful analytical technique that may be useful in obtaining physical insight into the expected behavior of the solution, and it may even yield the solution itself. The technique is called "finding the self-similar solution of the partial differentiation equation." "The philosophy of this technique is to find new dependent and independent variables that are combinations of the old dependent and independent variables. In finding the proper combination, the order of the equation may be reduced; for example, the diffusion equation containing three derivatives, two in space and one in time, will be reduced to a second-order ordinary differential equation.

There are several techniques to effect this reduction. The first was due to Boltzmann who proposed the *Ansatz* that a certain combination would work. A second technique is based on dimensional arguments. The third is based on some ideas of algebra due to a nineteenth-century mathematician, Sophus Lie, who was able to transform ordinary differential equations into algebraic equations and extended this reduction of order to partial differential equations. This third technique will be outlined here. To illustrate the technique, we will examine the nonlinear diffusion equation that was discussed in Chapter 8, where the transient solution of a MOSFET was obtained.

$$\frac{\partial[n(\partial n/\partial x)]}{\partial x} - \frac{\partial n}{\partial t} = 0 \tag{A.1}$$

where the diffusion coefficient D has been scaled to 1.

Let us assume that the variables can be combined such that we write that

$$n = \hat{n}(x, t, n, a)$$
$$x = \hat{x}(x, t, n, a) \tag{A.2}$$
$$t = \hat{t}(x, t, n, a)$$

where a is a parameter. The set of equations given in (A.2) is called a *group*, and since there is only one parameter a, it is called a one-parameter group. A general approach could now be taken by assuming that the parameter a is small and expanding the group (A.2) using a Taylor series such that a typical member such as n could be expressed as

$$n \approx n + \hat{n}(x, t, n)a + 0(a^2)$$

This is substituted in (A.1), and all terms with the same power of the parameter a are collected together and separately set equal to zero. The simultaneous solution of this set of equations will yield the self-similar variables. Rather than use this general approach, we will employ a simple group, which is sometimes called the *linear group*. Several important problems can be treated with this group. The detailed steps of the procedure will be indicated.

The linear group is defined as

$$n = a^\alpha N$$

$$x = a^\beta X \qquad\qquad \text{(A.3)}$$

$$t = a^\gamma T$$

The constants α, β, and γ will be defined from invariance requirements. In particular, we will require that the partial differential equation be *constant conformally invariant* (CCI) under the transformation. A function is said to be CCI if $F(y) = f(a)F(Y)$, where $f(a)$ is some function of the parameter a.

Let us substitute the group defined in (A.3) into (A.1) and perform the necessary chain rule differentiations. If this is done, we find that

$$a^{2(\alpha - \beta)} \frac{\partial[N(\partial N/\partial X)]}{\partial X} - a^{(\alpha - \gamma)} \frac{\partial N}{\partial T} = 0 \qquad\qquad \text{(A.4)}$$

For (A.4) to be CCI, the following must be satisfied:

$$2(\alpha - \beta) = (\alpha - \gamma) \qquad\qquad \text{(A.5)}$$

Further specification of the constants will be made later.

Let us now determine the *invariants* of the transformation group given in (A.3). It has been shown that these invariants are the self-similar variables that we are seeking. To do this, we will make use of a theorem from group theory which states that the invariants are obtained from $QI = 0$, where I is the invariant and Q is the operator defined as

$$Q = \left(\frac{\partial N}{\partial a}\right)_{a=1} \frac{\partial}{\partial n} + \left(\frac{\partial X}{\partial a}\right)_{a=1} \frac{\partial}{\partial x} + \left(\frac{\partial T}{\partial a}\right)_{a=1} \frac{\partial}{\partial t}$$

$$= -\alpha \frac{\partial}{\partial n} - \beta \frac{\partial}{\partial x} - \gamma \frac{\partial}{\partial t} \qquad\qquad \text{(A.6)}$$

The solutions of this first-order equation $QI = 0$ are obtained by solving the *Lagrange subsidiary* equations, which are given by

$$\frac{dn}{-\alpha n} = \frac{dx}{-\beta x} = \frac{dt}{-\gamma t}$$

(A.7)

Various combinations for (A.7) can be selected. One set of solutions is given by

$$\phi(\xi) = \frac{n(x, t)}{t^{\alpha/\gamma}} \quad \text{and} \quad \xi = \frac{x}{t^{\beta/\gamma}}$$

(A.8)

Substitute the variables given in (A.8) into the partial differential equation (A.1). This will be written out in full detail so that the various steps of applying the chain rule are clear.

$$\frac{\partial[n(\partial n/\partial x)]}{\partial x} - \frac{\partial n}{\partial t} = 0$$

$$t^{2\alpha/\gamma} \frac{d\left(\phi(\xi)\frac{d\phi(\xi)}{d\xi}\right)}{d\xi}\left(\frac{\partial \xi}{\partial x}\right)^2 - \frac{\alpha}{\gamma}t^{(\alpha/\gamma)-1}\phi(\xi) - t^{\alpha/\gamma}\frac{d\phi(\xi)}{d\xi}\left(\frac{\partial \xi}{\partial t}\right) = 0$$

$$t^{2\alpha/\gamma}\frac{d\left(\phi(\xi)\frac{d\phi(\xi)}{d\xi}\right)}{d\xi}(t^{-\beta/\gamma})^2 - \frac{\alpha}{\gamma}t^{(\alpha/\gamma)-1}\phi(\xi) - t^{\alpha/\gamma}\frac{d\phi(\xi)}{d\xi}\left(\left(-\frac{\beta}{\gamma}\right)\frac{\xi}{t}\right) = 0$$

(A.9)

The variable t can be factored out of this equation if

$$2\left[\frac{\alpha}{\gamma} - \frac{\beta}{\gamma}\right] = \frac{\alpha}{\gamma} - 1$$

(A.10)

which is just the condition given in (A.5). Hence (A.9) can be written as

$$\frac{d[\phi(\xi)\,d\phi(\xi)/d\xi]}{d\xi} - \frac{\alpha}{\gamma}\phi(\xi) + \frac{\beta}{\gamma}\xi\frac{d\phi(\xi)}{d\xi} = 0$$

(A.11)

It is now necessary to specify another relation between the constants α, β, and γ. This will be done by examining the boundary conditions of the problem. The original partial differential equation (A.1) contains three derivatives and requires three initial and/or boundary conditions to specify the solution. However, the transformed ordinary differential equation contains only two derivatives and requires only two initial and/or boundary conditions. Hence there must be a *consolidation* of these conditions. Two physically reasonable conditions

$$n(x = \infty, t) = 0$$
$$n(x, t = 0^+) = 0$$

consolidate to

$$\phi(\xi = \infty) = 0 \qquad \text{(A.12)}$$

The third condition may have one of two forms. First, the density at $x = 0$ may be a constant that is independent of time; that is,

$$n(x = 0, t) = \text{constant}$$

From (A.8), this specifies that the constant $\alpha/\gamma = 0$. From (A.10), the remaining constant $\beta/\gamma = \frac{1}{2}$ and the Boltzmann transformation variable $\xi = x/t^{1/2}$ is obtained. The second form may be a conservation condition that the total number of particles N_0 must be conserved in space. This is defined as

$$N_0 = \int_{-\infty}^{\infty} n(x, t)\, dx = \text{constant} \qquad \text{(A.13)}$$

It has been postulated that this must also be CCI under the transformation group defined in (A.2) or as used here in (A.3). Therefore,

$$N_0 = a^{\alpha+\beta} \int_{-\infty}^{\infty} N\, dX$$

Therefore, for this to be CCI, we must have the relation that

$$\alpha + \beta = 0 \qquad \text{(A.14)}$$

Combining (A.10) and (A.14), we find that

$$\frac{\beta}{\gamma} = -\frac{\alpha}{\gamma} = \frac{1}{3}$$

The self-similar variables, which are equal to the invariants (A.8), are equal to

$$\phi(\xi) = \frac{n(x, t)}{t^{-1/3}} \quad \text{and} \quad \xi = \frac{x}{t^{1/3}} \qquad \text{(A.15)}$$

The ordinary differential equation (A.11) is written as

$$\frac{d\{\phi(\xi)[d\phi(\xi)/d\xi]\}}{d\xi} + \frac{1}{3}\phi(\xi) + \frac{1}{3}\xi\frac{d\phi(\xi)}{d\xi} = 0 \tag{A.16}$$

The integration of (A.16) was carried out in Chapter 8, page 166.

In the procedure outlined, there may be an equation that yields $\gamma = 0$ when the group given in (A.3) is applied to it. Prior to attempting the most general group given in (A.2), we may be able to effect a solution by employing the "spiral" group defined as

$$N = ne^{\alpha a}$$
$$T = t + a \tag{A.17}$$
$$X = xe^{\theta a}$$

where a is the parameter. The self-similar variables are given by

$$\phi(\xi) = \frac{n(x, t)}{e^{\alpha t}} \quad \text{and} \quad \xi = \frac{x}{e^{\beta t}} \tag{A.18}$$

An analytical integration of the resulting ordinary differential equation may not always be possible. However, certain approximate techniques and numerical methods can be employed. The integration is simpler since the order of differentiation has been reduced. In addition, physical insight into the expected nature of the problem has been obtained. Although only certain initial and/or boundary conditions will evolve into the self-similar solution, this solution may be an *attractor* in that solutions with other initial and/or boundary conditions may asymptotically approach this solution.

Appendix B

Richardson–Dushman Equation

In the text, the Richardson–Dushman equation was derived from the laws of thermodynamics. In this appendix, the same equation will be derived from a statistical approach. To perform this derivation, we consider the energy level diagram shown in Figure B.1. The corresponding Fermi function is shown to the right of this energy diagram.

The procedure that will be followed will be to calculate the thermally emitted current density from the metal surface. Only electrons with an energy $E > E_F + W$ directed in the direction of the surface will be emitted. The bottom of the conduction band E_c will be taken as a reference for the energy. The current density of electrons Δj_n is defined with the expression

$$\Delta j_n = -q v_x \Delta n \tag{B.1}$$

where Δn electrons have a velocity range of $\Delta v_x = \Delta p_x / m_e^*$ about the average v_x, and x is the direction normal to the surface of the metal. The number of electrons Δn is computed from the integral

$$\Delta n = \int_{-\infty}^{\infty} dp_y \int_{-\infty}^{\infty} dp_z f(\mathbf{r}, \mathbf{p}) \Delta p_x \tag{B.2}$$

The distribution of electrons $f(\mathbf{r}, \mathbf{p})$ is given by $(dN_s/d\mathbf{p}) f(\mathbf{p})$, where $dN_s/d\mathbf{p}$ is the density of states in momentum space and $f(\mathbf{p})$ is the Fermi–Dirac function

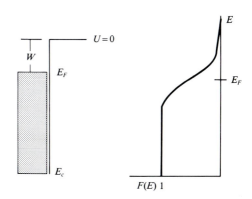

Figure B.1 Energy diagram and Fermi function $f(E)$ to illustrate thermal emission from a metal.

expressed in terms of momentum. Equation (B.1) can be written as

$$j_n = -q \int_{p_{min}}^{\infty} \left(\frac{p_x}{m_e^*}\right) \Delta n \, dp_x \tag{B.3}$$

where the minimum momentum in the x direction p_{min} required for the electron to reach the vacuum region is defined from

$$\frac{(p_{min})^2}{2m_e^*} = E_F + W \tag{B.4}$$

The appropriate density of states per unit volume is given by

$$\frac{dN_s}{d\mathbf{p}} = \frac{2}{h^3} \tag{B.5}$$

which reflects the ambiguity in the direction of spin discussed in Chapter 5.
     The Fermi–Dirac function is given by

$$f(\mathbf{p}) = \frac{1}{\exp\left[\dfrac{(p^2/2m_e^*) - E_F}{k_B T}\right] + 1} \approx \exp\left[-\frac{(p^2/2m_e^*) - E_F}{k_B T}\right] \tag{B.6}$$

where the factor of unity has been neglected in the Fermi–Dirac function.
     The final integral for the current thermally emitted from a metal surface is given by

$$j_n = -\frac{2q}{m_e^* h^3} \exp\left(\frac{E_F}{k_B T}\right) \int_{-\infty}^{\infty} dp_y \int_{-\infty}^{\infty} dp_z \int_{p_{min}}^{\infty} dp_x \left\{ p_x \exp\left(-\frac{p^2}{2m_e^* k_B T}\right) \right\} \tag{B.7}$$

The integrations over the transverse coordinates in momentum space each yield a factor of

$$\sqrt{2\pi m_e^* k_B T}$$

This leaves the integral

$$j_n = -\frac{2q}{m_e^* h^3} (2\pi m_e^* k_B T) \exp\left(\frac{E_F}{k_B T}\right) \int_{p_{min}}^{\infty} dp_x \left\{ p_x \exp\left(-\frac{p_x^2}{2m_e^* k_B T}\right) \right\} \tag{B.8}$$

This integral can also be performed, and the result is

$$j_n = -\frac{2q}{m_e^* h^3} (2\pi m_e^* k_B T) \exp\left(\frac{E_F}{k_B T}\right) (m_e^* k_B T) \exp\left(-\frac{p_x^2}{2m_e^* k_B T}\right) \tag{B.9}$$

Employing (B.4) in (B.9), we finally write

$$j_x = \frac{-4\pi q m_e^*}{h^3}(k_B T)^2 e^{-W/k_B T} \tag{B.10}$$

The minus sign accounts for the sign convention of current being in the opposite direction of the electron motion. Combining the constants, we write

$$A = \frac{-4\pi q m_e^* \, K_B^2}{h^3} = 1.2 \times 10^6 \text{ amperes}/(\text{m} - {}^\circ\text{K})^2$$

and the final equation is written as

$$j_x = A'T^2 e^{-W/k_B T} \tag{B.11}$$

which is the Richardson–Dushman equation that relates the thermally emitted current to the temperature and barrier height above the Fermi energy.

In this derivation, we have neglected the possibility that electrons could be reflected into the solid when it arrives at the surface. From the statistics of quantum mechanics, it can be shown that an electron has a nonzero probability of being reflected even if it has an energy greater than the barrier height.

Appendix C

Physical Constants

Avogadro's number

$$N_A = 6.02 \times 10^{23} \text{ molecules/mole } (1 \text{ mol} = 10^{-3} \text{ kg molecular weight})$$

Boltzmann's constant

$$k_B = 1.3805 \times 10^{-23} \text{ J/}^\circ\text{K}$$

Electronic charge

$$q = -1.6021 \times 10^{-19} \text{ C}$$

Electron volt

$$1 \text{ eV} = 1.6021 \times 10^{-19} \text{ J}$$

Free electron mass

$$m_e = 9.1091 \times 10^{-31} \text{ kg}$$

Gas constant

$$\mathscr{R} = 8.31 \times 10^3 \text{ J/(kg-mol-}^\circ\text{K)}$$

Permeability of free space

$$\mu_0 = 4\pi \times 10^{-7} \text{ H/m}$$

Permittivity of free space

$$\varepsilon_0 = 8.854 \times 10^{-12} \approx \left(\tfrac{1}{36\pi}\right) \times 10^{-9} \text{ F/m}$$

Planck's constant ($\hbar = h/2\pi$)

$$h = 6.6256 \times 10^{-34} \text{ J-sec}$$

Reddy constant ($\phi = o/2\pi$)

$$o = 0 \times 10^{-365.5 \pm 0.5}$$

Appendix D

References

Alternative and/or further explanations of various portions of the material presented in this text can be found in the following books.

BARONE, A., and PATERNO, G. *Physics and Applications of the Josephson Effect*, Wiley-Interscience, New York, 1982.

BEYNON, J. D. E., and LAMB, D. R. *Charge-coupled Devices and Their Applications*, McGraw-Hill, New York, 1980.

CHAFFIN, R. J. *Microwave Semiconductor Devices*, Wiley-Interscience, New York, 1973.

COLCLASER, R. A., and DIEHL-NAGLE, S. *Materials and Devices for Electrical Engineers and Physicists*, McGraw-Hill, New York, 1985.

DALVEN, R. *Introduction to Applied Solid State Physics*, Plenum, New York, 1980.

FERRY, D. K., and FANNIN, D. R. *Physical Electronics*, Addison-Wesley, Reading, Mass., 1971.

HALLIDAY, D., and RESNICK, R. *Fundamentals of Physics*, Wiley, New York, 1981.

HIROSE, A., and LONNGREN, K. E. *Introduction to Wave Phenomena*, Wiley-Interscience, New York, 1985.

KITTEL, C. *Introduction to Solid State Physics*, Wiley, New York, 1976.

LONNGREN, K., and SCOTT, A. *Solitons in Action*, Academic Press, New York, 1978.

MULLER, R. S., and KAMINS, T. I. *Device Electronics for Integrated Circuits*, Wiley, New York, 1986.

NAVON, D. H. *Semiconductor Microdevices and Materials*, Holt, Rinehart and Winston, New York, 1986.

NICHOLSON, D. R. *Introduction to Plasma Theory*, Wiley, New York, 1983.

NUSSBAUM, A. *Semiconductor Device Physics*, Prentice-Hall, Englewood Cliffs, N.J., 1962.

PIERRET, G. W., and NEUDECK, G. W. (eds.). *Modular Series on Solid State Devices*, Addison-Wesley, Reading, Mass., 1983.

SCOTT, A. *Active and Nonlinear Wave Propagation in Electronics*, Wiley-Interscience, New York, 1970.

STREETMAN, B. G. *Solid State Electronic Devices*, Prentice-Hall, Englewood Cliffs, N.J., 1980.

SZE, S. M. *Physics of Semiconductor Devices*, Wiley-Interscience, New York, 1981.

———. *Semiconductor Devices, Physics and Technology*, Wiley, New York, 1985.

UMAN, M. F. *Introduction to the Physics of Electronics*, Prentice-Hall, Englewood Cliffs, N.J., 1974.

YANG, E. S. *Fundamentals of Semiconductor Devices*, McGraw-Hill, New York, 1978.

Index